TELECOMMUNICATIONS AND MEDIA ISSUES

TELECOMMUNICATIONS AND MEDIA ISSUES

ALANE N. MOLLER AND CHARLES E. PLETSON
EDITORS

Nova Science Publishers, Inc.
New York

For permission to use material from this book please contact us:
Telephone 631-231-7269; Fax 631-231-8175
Web Site: http://www.novapublishers.com

NOTICE TO THE READER

The Publisher has taken reasonable care in the preparation of this book, but makes no expressed or implied warranty of any kind and assumes no responsibility for any errors or omissions. No liability is assumed for incidental or consequential damages in connection with or arising out of information contained in this book. The Publisher shall not be liable for any special, consequential, or exemplary damages resulting, in whole or in part, from the readers' use of, or reliance upon, this material.

Independent verification should be sought for any data, advice or recommendations contained in this book. In addition, no responsibility is assumed by the publisher for any injury and/or damage to persons or property arising from any methods, products, instructions, ideas or otherwise contained in this publication.

This publication is designed to provide accurate and authoritative information with regard to the subject matter covered herein. It is sold with the clear understanding that the Publisher is not engaged in rendering legal or any other professional services. If legal or any other expert assistance is required, the services of a competent person should be sought. FROM A DECLARATION OF PARTICIPANTS JOINTLY ADOPTED BY A COMMITTEE OF THE AMERICAN BAR ASSOCIATION AND A COMMITTEE OF PUBLISHERS.

LIBRARY OF CONGRESS CATALOGING-IN-PUBLICATION DATA

Telecommunications and media issues / Alane N. Moller and Charles E. Pletson (editor).
  p. cm.
 ISBN 978-1-60456-294-1 (hardcover)
 1. Television broadcasting--United States. 2. Television broadcasting--Law and legislation--United States. I. Moller, Alane N. II. Pletson, Charles E.
PN1992.3.U5T32 2008
302.23--dc22
                        2007050783

Published by Nova Science Publishers, Inc. ✢ New York

CONTENTS

PREFACE

In a society predicated on information, the media has a pervasive presence. From government policy to leisure television, the information age touches us all. The papers collected in this book constitute some of today's leading analyses of the information industry. Together, these essays represent a needed foundation for understanding the present state and future development of the mass media. Current trends in telecommunications as well as media impact on public opinion are presented.

Chapter 1 - Two prominent television events placed increased attention on the Federal Communications Commission (FCC) and the broadcast indecency statute that it enforces. The airing of an expletive by Bono during the 2003 Golden Globe Awards, as well as the "wardrobe malfunction" that occurred during the 2004 Super Bowl half-time show, gave broadcast indecency prominence in the 109th and 110th Congresses, and resulted in the enactment of P.L.109-235 (2006), which increased the penalties for broadcast indecency by tenfold.

Federal law makes it a crime to utter "any obscene, indecent, or profane language by means of radio communication" (18 U.S.C. § 1464). Violators of this statute are subject to fines and imprisonment of up to two years, and the FCC may enforce this provision by forfeiture or revocation of a broadcaster's license. The FCC has found that, for material to be "indecent," it "must describe or depict sexual or excretory organs or activities," and "must be patently offensive as measured by contemporary community standards for the broadcast medium." The federal government's authority to regulate material that is "indecent" but not obscene was upheld by the Supreme Court in *Federal Communications Commission v. Pacifica Foundation*, which found that prohibiting such material during certain times of the day does not violate the First Amendment.

In 1992, Congress enacted P.L. 102-356 (47 U.S.C. § 303 note), section 16(a) of which, as interpreted by the courts, requires the FCC to prohibit "indecent" material on broadcast radio and broadcast television from 6 a.m. to 10 p.m. Under P.L. 109-235, "indecent" broadcasts are now subject to a fine of up to "$325,000 for each violation or each day of continuing violation, except that the amount assessed for any continuing violation shall not exceed a total of $3,000,000 for any single act or failure to act." Fines may be levied against broadcast stations, but not against broadcast networks. The FCC appears to have the statutory authority to fine performers as well (up to $32,500 per incident), but has taken the position that "[c]ompliance with federal broadcast decency restrictions is the responsibility of the station that chooses to air the programming, not the performers."

The federal restriction on "indecent" material applies only to broadcast media, and this stems from the fact that there are a limited number of broadcast frequencies available and that the Supreme Court, therefore, allows the government to regulate broadcast media more than other media. This report discusses the legal evolution of the FCC's indecency regulations, and provides an overview of how the current regulations have been applied. The final section of the report considers whether prohibiting the broadcast of "indecent" words regardless of context would violate the First Amendment. This question arises because the Supreme Court in *Pacifica* left open the question whether broadcasting an occasional expletive, as in the Bono case, would justify a sanction.

Chapter 2 - Spam, also called unsolicited commercial email (UCE) or "junk email," aggravates many computer users. Not only can spam be a nuisance, but its cost may be passed on to consumers through higher charges from Internet service providers who must upgrade their systems to handle the traffic. Also, some spam involves fraud, or includes adult-oriented material that offends recipients or that parents want to protect their children from seeing. Proponents of UCE insist it is a legitimate marketing technique that is protected by the First Amendment, and that some consumers want to receive such solicitations.

On December 16, 2003, President Bush signed into law the Controlling the Assault of Non-Solicited Pornography and Marketing (CAN-SPAM) Act, P.L. 108-187. It went into effect on January 1, 2004. The CAN-SPAM Act does not ban UCE. Rather, it allows marketers to send commercial email as long as it conforms with the law, such as including a legitimate opportunity for consumers to "opt-out" of receiving future commercial emails from that sender. It preempts state laws that specifically address spam, but not state laws that are not specific to email, such as trespass, contract, or tort law, or other state laws to the extent they relate to fraud or computer crime. It does not require a centralized "Do Not Email" registry to be created by the Federal Trade Commission (FTC), similar to the National Do Not Call registry for telemarketing. The law requires only that the FTC develop a plan and timetable for establishing such a registry, and to inform Congress of any concerns it has with regard to establishing it. The FTC submitted a report to Congress on June 15, 2004, concluding that a Do Not Email registry could actually increase spam.

Proponents of CAN-SPAM have argued that consumers are most irritated by *fraudulent* email, and that the law should reduce the volume of such email because of the civil and criminal penalties included therein. Opponents counter that consumers object to *unsolicited* commercial email, and since the law legitimizes commercial email (as long as it conforms with the law's provisions), consumers actually may receive more, not fewer, UCE messages. Thus, whether or not "spam" is reduced depends in part on whether it is defined as only fraudulent commercial email, or all unsolicited commercial email. Many observers caution that consumers should not expect any law to solve the spam problem — that consumer education and technological advancements also are needed.

Note: This report was originally written by Marcia S. Smith; the author acknowledges her contribution to CRS coverage of this issue area.

Chapter 3 - Various federal officials have spoken in favor of extending the Federal Communication Commission's indecency restriction, which currently applies to broadcast television and radio, to cable and satellite television. This report examines whether such an extension would violate the First Amendment's guarantee of freedom of speech.

The FCC's indecency restriction was enacted pursuant to a federal statute that, insofar as it was found constitutional, requires the FCC to promulgate regulations to prohibit the broadcast of indecent programming from 6 a.m. to 10 p.m. The FCC has found that, for material to be "indecent," it "must describe or depict sexual or excretory organs or activities," and "must be patently offensive as measured by contemporary community standards for the broadcast medium."

In 1978, in *Pacifica*, the Supreme Court held that, because broadcast radio and television have a "uniquely pervasive presence" and are "uniquely accessible to children," the government may, during certain times of day, prohibit "[p]atently offensive, indecent material" on these media. In 1996, however, a Supreme Court plurality held that, with respect to "how pervasive and intrusive [television] programming is . . . cable and broadcast television differ little, if at all."

Then, in 2000, the Court held that governmental restrictions on speech on cable television are, unlike those on broadcast media, entitled to strict scrutiny. Thus, whereas, in *Pacifica*, the Court upheld a restriction on "indecent" material on broadcast media without applying strict scrutiny, the Court apparently would not uphold a comparable restriction on "indecent" material on cable television unless the restriction served a compelling governmental interest by the least restrictive means.

It seems uncertain whether the Court would find that denying minors access to "indecent" material on cable television would constitute a compelling governmental interest. Assuming that it would, then, whether or not there is a less restrictive means than a 6 a.m.-to-10 p.m. ban by which to deny minors access to "indecent" material on cable television, it appears that a strong case may be made that applying the FCC's indecency restrictions to cable television would violate the First Amendment. This is because, as the Supreme Court wrote when it struck down the ban on "indecent" material on the Internet, "the Government may not 'reduc[e] the adult population . . . to . . . only what is fit for children.'" In addition, the Court, in the 2000 case mentioned above, struck down a speech restriction on cable television, in part because "for two-thirds of the day no household in those service areas could receive the programming, whether or not the household or the viewer wanted to do so."

Chapter 4 - The Deficit Reduction Act of 2005 (P.L. 109-171) directs that on February 18, 2009, over-the-air television broadcasts — which are currently provided by television stations in both analog and digital formats — will become digital only. Digital television (DTV) technology allows a broadcaster to offer a single program stream of high definition television (HDTV), or alternatively, multiple video program streams (multicasts). Households with over-the-air analog-only televisions will *no longer be able to receive television service* unless they either: (1) buy a digital-to-analog converter box to hook up to their analog television set; (2) acquire a digital television or an analog television equipped with a digital tuner; or (3) subscribe to cable, satellite, or telephone company television services, which will likely provide for the conversion of digital signals to their analog customers.

Households using analog televisions for viewing over-the-air television broadcasts are likely to be most affected by the digital transition. Of particular concern to many policymakers are low-income, elderly, disabled, non-English speaking, and minority populations. Many of these groups tend to rely more on over-the-air television, and are thus more likely impacted by the digital transition.

The Deficit Reduction Act of 2005 established a digital-to-analog converter box program — administered by the National Telecommunications and Information Administration (NTIA) of the Department of Commerce — that will partially subsidize consumer purchases of converter boxes. NTIA will provide up to two forty-dollar coupons to requesting U.S.

households. The coupons are to be issued between January 1, 2008, and March 31, 2009, and must be used within three months after issuance towards the purchase of a stand-alone device used solely for digital-to-analog conversion.

The preeminent issue for Congress is ensuring that American households are prepared for the February 17, 2009 DTV transition deadline, thereby minimizing a scenario whereby television sets across the nation "go dark." Specifically, Congress is actively overseeing the activities of federal agencies responsible for the digital transition — principally the Federal Communications Commission (FCC) and the NTIA — while assessing whether additional federal efforts are necessary, particularly with respect to public education and outreach. The Congress is also monitoring the extent to which private sector stakeholders take appropriate and sufficient steps to educate the public and ensure that all Americans are prepared for the digital transition. DTV- related bills, which address public education (H.R. 608, H.R. 2566, H.R. 2917, and S. 2125), have been introduced into the 110th Congress. At issue is whether the federal government's current programs and reliance on private sector stakeholders will lead to a successful digital transition with a minimum amount of disruption to American TV households or, alternatively, whether additional legislative measures are warranted.

This report will be updated as events warrant.

Chapter 5 - To assist parents in supervising the television viewing habits of their children, the Communications Act of 1934 (as amended by the Telecommunications Act of 1996) requires that, as of January 1, 2000, new television sets with screens 13 inches or larger sold in the United States be equipped with a "V-chip" to control access to programming that parents find objectionable. Use of the V-chip is optional. In March 1998, the Federal Communications Commission (FCC) adopted the industry-developed ratings system to be used in conjunction with the V-chip. Congress and the FCC have continued monitoring implementation of the V-chip. Some are concerned that it is not effective in curbing the amount of TV violence viewed by children and want further legislation.

In July 2004, the FCC initiated a Notice of Inquiry (NOI) to seek comments relating to the "presentation of violent programing and its impact on children." The Report in this proceeding was released by the FCC on April 25, 2007. In the report, the FCC, among other findings, (1) found that on balance, research provides strong evidence that exposure to violence in the media can increase aggressive behavior in children, at least in the short term; (2) stated that the V-chip is of limited effectiveness in protecting children from violent television content and observed that cable operator-provided advanced parental controls do not appear to be available on a sufficient number of cable-connected television sets to be considered an effective solution at this time; and (3) found that studies and surveys demonstrate that the voluntary TV ratings system is of limited effectiveness in protecting children from violent television content.

Congress may wish to consider a number of possible options to support parents in controlling their children's access to certain programming. Some of these options would require only further educational outreach to parents, while others would require at least regulatory, if not legislative, action. Specifically, Congress may wish to consider ways to promote awareness of the V-chip and the ratings system; whether the current set of media-specific ratings will remain viable in the future or whether a uniform system would better serve the needs of consumers; and whether independent ratings systems and an "open" V-chip that would allow consumers to select the ratings systems they use would be more appropriate than the current system.

Chapter 6 - In November 2003, the Federal Communications Commission (FCC) adopted a rule that required all digital devices capable of receiving digital television (DTV) broadcasts over the air, and sold after July 1, 2005, to incorporate technology that would recognize and abide by the broadcast video flag, a content-protection signal that broadcasters may choose to embed into a digital broadcast transmission as a way to prevent unauthorized redistribution of DTV content. However, in October 2004, the American Library Association and eight organizations representing a large number of libraries and consumers filed a lawsuit that challenged the power of the FCC to promulgate such a rule. In May 2005, the United States Court of Appeals for the District of Columbia Circuit ruled in *American Library Association v. Federal Communications Commission* that the FCC had exceeded the scope of its delegated authority in imposing the broadcast flag regime, and the court thus reversed and vacated the FCC's broadcast flag order.

Parties holding a copyright interest in content transmitted through DTV broadcasts, in particular broadcasters and television program creators, remain concerned about the unauthorized distribution and reproduction of copyrighted DTV content and thus continue to advocate the adoption of a broadcast video flag. However, several consumer, educational, and technology groups raise objections to the broadcast flag because, in their view, it would place technological, financial, and regulatory burdens that may stifle innovation, limit the consumer's ability to use DTV broadcasts in accordance with the Copyright Act's "fair use" principles, and possibly frustrate the use of digital television content by educators and librarians in distance education programs.

This report provides a brief explanation of the broadcast video flag and its relationship to digital television and summarizes the *American Library Association* judicial opinion. The report also examines a legislative proposal introduced in the 109th Congress, the Digital Content Protection Act of 2006, which appeared as portions of two bills, S. 2686 and H.R. 5252 (as reported in the Senate), that would have expressly granted statutory authority to the FCC under the Communications Act of 1934 to promulgate regulations implementing a broadcast video flag system. Although not enacted, these bills represent approaches to authorizing the broadcast video flag system that may be of interest to the 110th Congress.

Chapter 7 - Because existing international agreements relevant to broadcasting protections do not cover advancements in broadcasting technology that were not envisioned when they were concluded, in 1998 the Standing Committee on Copyright and Related Rights (SCCR) of the World Intellectual Property Organization (WIPO) decided to proceed with efforts to negotiate and draft a new treaty that would extend protection to new methods of broadcasting, but has yet to achieve consensus on a text. In recent years, a growing signal piracy problem has increased the urgency of concluding a new treaty, resulting in a decision to restrict the focus to signal-based protections for traditional broadcasting organizations and cablecasting. Consideration of controversial issues of protections for webcasting (advocated by the United States) and simulcasting will be postponed. However, considerable work remains to achieve a final proposed text as the basis for formal negotiations to conclude a treaty by the end of 2007, as projected. A concluded treaty would not take effect for the United States unless Congress enacts implementing legislation and the United States ratifies the treaty with the advice and consent of the Senate. Noting that the United States is not a party to the 1961 Rome Convention, various stakeholders have argued that a new broadcasting treaty is not needed, that any new treaty should not inhibit technological innovation or consumer use, and that Congress should exercise greater oversight over U.S. participation in the negotiations.

Chapter 8 - Protecting audio content broadcasted by digital and satellite radios from unauthorized dissemination and reproduction is a priority for producers and owners of those copyrighted works. One technological measure that has been discussed is the Audio Protection Flag (APF or "audio flag"). The audio flag is a special signal that would be imbedded into digital audio radio transmissions, permitting only authorized devices to play back copyrighted audio transmissions or allowing only limited copying and retention of the content. Several bills introduced in the 109th Congress would have granted the Federal Communications Commission (FCC) authority to promulgate regulations to implement the audio flag. The parties most likely affected by any audio flag regime (including music copyright owners, digital radio broadcasters, stereo equipment manufacturers, and consumers) are divided as to the anticipated degree and scope of the impact that a government-mandated copyright protection scheme would have on the "fair use" rights of consumers to engage in private, noncommercial home recording. Critics of the audio flag proposal are concerned about its effect on technological innovation. However, proponents of the audio flag feel that such digital rights management (DRM) technology is needed to thwart piracy or infringement of intellectual property rights in music, sports commentary and coverage, and other types of copyrighted content that is transmitted to the public by emerging high-definition digital radio services (HD Radio) and satellite radio broadcasters.

This report provides a brief explanation of the audio flag and its relationship to digital audio radio broadcasts, and summarizes legislative proposals considered by the 109th Congress, including H.R. 4861 (Audio Broadcast Flag Licensing Act of 2006) and S. 2686 (Digital Content Protection Act of 2006), that would have authorized its adoption. Although not enacted, these two bills represent approaches that may be taken in the 110th Congress to authorize the use of an audio flag for protecting broadcast digital audio content.

In: Telecommunications and Media Issues
Editors: A. N. Moller and C. E. Pletson, pp. 1-31

ISBN: 978-1-60456-294-1
© 2008 Nova Science Publishers, Inc.

.

Chapter 1

REGULATION OF BROADCAST INDECENCY: BACKGROUND AND LEGAL ANALYSIS[*]

Henry Cohen and Kathleen Ann Ruane

ABSTRACT

Two prominent television events placed increased attention on the Federal Communications Commission (FCC) and the broadcast indecency statute that it enforces. The airing of an expletive by Bono during the 2003 Golden Globe Awards, as well as the "wardrobe malfunction" that occurred during the 2004 Super Bowl half-time show, gave broadcast indecency prominence in the 109th and 110th Congresses, and resulted in the enactment of P.L.109-235 (2006), which increased the penalties for broadcast indecency by tenfold.

Federal law makes it a crime to utter "any obscene, indecent, or profane language by means of radio communication" (18 U.S.C. § 1464). Violators of this statute are subject to fines and imprisonment of up to two years, and the FCC may enforce this provision by forfeiture or revocation of a broadcaster's license. The FCC has found that, for material to be "indecent," it "must describe or depict sexual or excretory organs or activities," and "must be patently offensive as measured by contemporary community standards for the broadcast medium." The federal government's authority to regulate material that is "indecent" but not obscene was upheld by the Supreme Court in *Federal Communications Commission v. Pacifica Foundation*, which found that prohibiting such material during certain times of the day does not violate the First Amendment.

In 1992, Congress enacted P.L. 102-356 (47 U.S.C. § 303 note), section 16(a) of which, as interpreted by the courts, requires the FCC to prohibit "indecent" material on broadcast radio and broadcast television from 6 a.m. to 10 p.m. Under P.L. 109-235, "indecent" broadcasts are now subject to a fine of up to "$325,000 for each violation or each day of continuing violation, except that the amount assessed for any continuing violation shall not exceed a total of $3,000,000 for any single act or failure to act." Fines may be levied against broadcast stations, but not against broadcast networks. The FCC appears to have the statutory authority to fine performers as well (up to $32,500 per incident), but has taken the position that "[c]ompliance with federal broadcast decency

---

[*] Excerpted from CRS Report RL32222, dated September 13, 2007.

restrictions is the responsibility of the station that chooses to air the programming, not the performers."

The federal restriction on "indecent" material applies only to broadcast media, and this stems from the fact that there are a limited number of broadcast frequencies available and that the Supreme Court, therefore, allows the government to regulate broadcast media more than other media. This report discusses the legal evolution of the FCC's indecency regulations, and provides an overview of how the current regulations have been applied. The final section of the report considers whether prohibiting the broadcast of "indecent" words regardless of context would violate the First Amendment. This question arises because the Supreme Court in *Pacifica* left open the question whether broadcasting an occasional expletive, as in the Bono case, would justify a sanction.

INTRODUCTION

Two prominent television events placed increased attention on the Federal Communications Commission (FCC) and the broadcast indecency statute that it enforces.[1] The airing of an expletive by Bono during the 2003 Golden Globe Awards, as well as the "wardrobe malfunction" that occurred during the 2004 Super Bowl half-time show, gave broadcast indecency prominence in the 109th and 110th Congresses, and resulted in the enactment of P.L.109-235 (2006), which increased the penalties for broadcast indecency by tenfold.

Federal law makes it a crime to utter "any obscene, indecent, or profane language by means of radio communication" (18 U.S.C. § 1464). Violators of this statute are subject to fines and imprisonment of up to two years, and the FCC may enforce this provision by forfeiture or revocation of a broadcaster's license. The FCC has found that, for material to be "indecent," it "must describe or depict sexual or excretory organs or activities," and "must be patently offensive as measured by contemporary community standards for the broadcast medium." The federal government's authority to regulate material that is "indecent" but not obscene was upheld by the Supreme Court in *Federal Communications Commission v. Pacifica Foundation*, which found that prohibiting such material during certain times of the day does not violate the First Amendment.

In 1992, Congress enacted P.L. 102-356 (47 U.S.C. § 303 note), section 16(a) of which, as interpreted by the courts, requires the FCC to prohibit "indecent" material on broadcast radio and broadcast television from 6 a.m. to 10 p.m. Under P.L. 109-235, "indecent" broadcasts are now subject to a fine of up to "$325,000 for each violation or each day of continuing violation, except that the amount assessed for any continuing violation shall not exceed a total of $3,000,000 for any single act or failure to act." Fines may be levied against broadcast stations, but not against broadcast networks. The FCC appears to have the statutory authority to fine performers as well (up to $32,500 per incident), but has taken the position that "[c]ompliance with federal broadcast decency restrictions is the responsibility of the station that chooses to air the programming, not the performers."

The federal restriction on "indecent" material applies only to broadcast media, and this stems from the fact that there are a limited number of broadcast frequencies available and that the Supreme Court, therefore, allows the government to regulate broadcast media more than other media. It appears likely that a court would find that to apply the FCC's indecency restriction to cable or satellite media would violate the First Amendment.[2]

This report discusses the evolution of the FCC's indecency regulations, provides an overview of how the current regulations have been applied, and examines indecency legislation that was introduced in the 109th and 110th Congress. (The bill that increased penalties is the only such legislation that was enacted.) The final section of this report considers whether prohibiting the broadcast of "indecent" words regardless of context would violate the First Amendment. This issue arises because the Supreme Court in *Pacifica* left open the question of whether broadcasting an occasional expletive, as in the Bono case, would justify a sanction.

BACKGROUND

On January 19, 2003, broadcast television stations in various parts of the country aired the Golden Globe Awards. During the awards, the singer Bono, in response to winning an award, said, "this is really, really f[***]ing brilliant."[3] In response to this utterance, the FCC received over 230 complaints alleging that the program was obscene or indecent, and requesting that the Commission impose sanctions on the licensees for the broadcast of the material in question.[4]

The Enforcement Bureau of the FCC issued a Memorandum Opinion and Order on October 3, 2003, denying the complaints and finding that the broadcast of the Golden Globe Awards including Bono's utterance did not violate federal restrictions regarding the broadcast of obscene and indecent material.[5] The Bureau dismissed the complaints primarily because the language in question did not describe or depict sexual or excretory activities or organs. The Bureau noted that while "the word 'f[***]ing' may be crude and offensive," it "did not describe sexual or excretory organs or activities. Rather, the performer used the word 'f[***]ing' as an adjective or expletive to emphasize an exclamation."[6] The Bureau added that in similar circumstances it "found that offensive language used as an insult rather than as a description of sexual or excretory activity or organs is not within the scope of the Commission's prohibition on indecent program content."[7]

The decision of the Enforcement Bureau was met with opposition from a number of organizations and Members of Congress, and an appeal was filed for review by the full Commission. FCC Chairman Michael Powell asked the full Commission to overturn the Enforcement Bureau's ruling.[8]

On March 18, 2004, the full Commission issued a *Memorandum Opinion and Order* granting the application for review and reversing the Enforcement Bureau's earlier opinion.[9] The Commission found that the broadcasts of the Golden Globe Awards violated 18 U.S.C. 1464, but declined to impose a forfeiture on the broadcast licensees because the Order reverses Commission precedent regarding the broadcast of the "F-word." This decision is discussed in detail below.

On February 1, 2004, CBS aired Super Bowl XXXVIII, with a half-time show produced by the MTV network. The show included performers singing and dancing provocatively, and ended with the exposure of the breast of one female performer. The network received numerous complaints regarding the half-time performance and FCC Chairman Michael Powell initiated a formal investigation into the incident.[10]

On September 22, 2004, the FCC released a *Notice of Apparent Liability for Forfeiture* finding that the airing of the Super Bowl halftime show "apparently violate[d] the federal restrictions regarding the broadcast of indecent material."[11] The *NAL* imposes a forfeiture in the aggregate amount of $550,000 on Viacom Inc., the licensee or ultimate parent of the licensees with regard to whom the complaint was filed.[12] On March 15, 2006, the FCC issued a *Forfeiture Order* imposing a mandatory forfeiture in the amount of $550,000 on CBS for the airing of the 2004 Super Bowl halftime show. CBS appealed to the U.S. Court of Appeals for the Third Circuit, which heard oral arguments in the case on September 11, 2007. This case is discussed in greater detail below.

EVOLUTION OF THE FCC'S INDECENCY REGULATIONS

Title 18 of the United States Code makes it unlawful to utter "any obscene, indecent, or profane language by means of radio communication."[13] Violators of this provision are subject to fines or imprisonment of up to two years. The Federal Communications Commission has the authority to enforce this provision by forfeiture or revocation of license.[14] The Commission's authority to regulate material that is indecent, but not obscene, was upheld by the Supreme Court in *Federal Communications Commission v. Pacifica Foundation*.[15] In *Pacifica*, the Supreme Court affirmed the Commission's order regarding the airing of comedian George Carlin's "Filthy Words" monologue.[16] In that order, the Commission determined that the airing of the monologue, which contained certain words that "depicted sexual and excretory activities in a patently offensive manner," at a time "when children were undoubtedly in the audience" was indecent and prohibited by 18 U.S.C. § 1464.[17] Pursuant to the Court's decision, whether any such material is "patently offensive" is determined by "contemporary community standards for the broadcast medium."[18] The Court noted that indecency is "largely a function of context — it cannot be judged in the abstract."[19]

The Commission's order in the *Pacifica* case relied partially on a spectrum scarcity argument; i.e., that there is a scarcity of spectrum space so the government must license the use of such space in the public interest, and partially on "principles analogous to those found in the law of nuisance."[20] The Commission noted that public nuisance law generally aims to channel the offensive behavior rather than to prohibit it outright. For example, in the context of broadcast material, channeling would involve airing potentially offensive material at times when children are less likely to be in the audience. In 1987, the Commission rejected the spectrum scarcity argument as a sufficient basis for its regulation of broadcast indecency, but noted that it would continue to rely upon the validity of the public nuisance rationale, including channeling of potentially objectionable material.[21] However, in its 1987 order, the Commission also stated that channeling based on a specific time of day was no longer a sufficient means to ensure that children were not in the audience when indecent material aired and warned licensees that indecent material aired after 10 p.m. would be actionable.[22] The Commission further clarified its earlier *Pacifica* order, noting that indecent language was not strictly limited to the seven words at issue in the original broadcast in question, and that repeated use of those words was not necessary to find that material in question was indecent.[23]

The Commission's 1987 orders were challenged by parties alleging that the Commission had changed its indecency standard and that the new standard was unconstitutional. In *Action for Children's Television v. Federal Communications Commission (ACT I)*, the United States Court of Appeals for the District of Columbia Circuit upheld the standard used by the Commission to determine whether broadcast material was indecent, but it vacated the Commission's order with respect to the channeling of indecent material for redetermination "after a full and fair hearing of the times at which indecent material may be broadcast."[24]

Following the court's decision in *Action for Children's Television (ACT I)*, a rider to the Commerce, Justice, State FY89 Appropriations Act required the FCC to promulgate regulations to ban indecent broadcasts 24 hours a day.[25] The Commission followed the congressional mandate and promulgated regulations prohibiting all broadcasts of indecent material.[26] The new regulations were challenged, and the United States Court of Appeals for the District of Columbia Circuit vacated the Commission's order.[27] In so doing, the court noted that in *ACT I* it held that Commission "must identify some reasonable period of time during which indecent material may be broadcast," thus precluding a ban on such broadcasts at all times.[28]

In 1992, Congress enacted the Public Telecommunications Act of 1992, which required the FCC to promulgate regulations to prohibit the broadcasting of indecent material from 6 a.m. to midnight, except for broadcasts by public radio and television stations that go off the air at or before midnight, in which case such stations may broadcast indecent material beginning at 10 p.m.[29] The Commission promulgated regulations as mandated in the act.[30] The new regulations were challenged, and a three-judge panel of the United States Court of Appeals for the District of Columbia Circuit subsequently vacated the Commission's order implementing the act and held the underlying statute unconstitutional.[31] In its order implementing the act, the FCC set forth three goals to justify the regulations: (1) ensuring that parents have an opportunity to supervise their children's listening and viewing of over-the-air broadcasts; (2) ensuring the well being of minors regardless of supervision; and (3) protecting the right of all members of the public to be free of indecent material in the privacy of their homes.[32] The court rejected the third justification as "insufficient to support a restriction on the broadcasting of constitutionally protected indecent material," but accepted the first two as compelling interests.[33] Despite the finding of compelling interests in the first two, the court found that both Congress and the FCC had failed "to tailor their efforts to advance these interests in a sufficiently narrow way to meet constitutional standards."[34]

Following the decision of the three-judge panel, the Commission requested a rehearing *en banc*.[35] The case was reheard on October 19, 1994, and, on June 30, 1995, the full court of appeals held the statute unconstitutional insofar as it prohibited the broadcast of indecent material between the hours of 10 p.m. and midnight on nonpublic stations.[36] In so doing, the court held that while the channeling of indecent broadcasts between midnight and 6 a.m. "would not unduly burden the First Amendment," the distinction drawn by Congress between public and non-public broadcasters "bears no apparent relationship to the compelling government interests that [the restrictions] are intended to serve."[37] The court remanded the regulations to the FCC with instructions to modify the regulations to permit the broadcast of indecent material on all stations between 10 p.m and 6 a.m.

Current Regulations

Following the decision in *ACT III*, the Commission modified its indecency regulations to prohibit indecent broadcasts from 6 a.m. to 10 p.m.[38] The modified regulations became effective August 28, 1995.[39] These regulations have been enforced primarily with respect to radio broadcasts and thus have been applied more often to indecent language rather than to images.[40] Under these regulations, broadcasts deemed indecent were subject to a forfeiture of up to $32,500 per violation,[41] with the FCC's considering each utterance of an indecent word as a separate violation, rather than viewing the entire program as a single violation.[42]

Fines may be levied against broadcast stations, but not against broadcast networks. The FCC appears also to have the statutory authority to fine performers for uttering indecent words,[43] but it has taken the position that "[c]ompliance with federal broadcast decency restrictions is the responsibility of the station that chooses to air the programming, not the performers."[44]

On June 15, 2006, the President signed S. 193, 109th Congress, into law, and it became P.L. 109-235, the Broadcast Decency Enforcement Act of 2005. This law increased the penalty for indecent broadcasts tenfold, to $325,000 for each violation, with a maximum of $3 million "for any single act or failure to act." This increased penalty may be levied against "a broadcast station licensee or permittee; or . . . an applicant for any broadcast license, permit, certificate, or other instrument or authorization issued by the Commission." If the FCC were to change its policy and impose fines on performers, it could apparently do so only under the provision (which remains in effect) that authorizes forfeitures of up to $32,500 per violation.[45]

To determine whether broadcast material is in fact indecent, the Commission must make two fundamental determinations: (1) that the material alleged to be indecent falls within the subject matter scope of the definition of indecency — the material in question must describe or depict sexual or excretory organs or activities; and (2) that the broadcast is patently offensive as measured by contemporary community standards for the broadcast medium.[46] If the material in question does not fall within the subject matter scope of the indecency definition, or if the broadcast occurred during the "safe harbor" hours (between 10 p.m. and 6 a.m.), the complaint is usually dismissed. However, if the Commission determines that the complaint meets the subject matter requirements and was aired outside the "safe harbor" hours, the broadcast in question is evaluated for patent offensiveness. The Commission notes that in determining whether material is patently offensive, the full context is very important, and that such determinations are highly fact-specific.

The Commission has identified three factors that have been significant in recent decisions in determining whether broadcast material is patently offensive:

1. the explicitness or graphic nature of the description or depiction of sexual or excretory organs or activities;
2. whether the material dwells on or repeats at length descriptions of sexual or excretory organs or activities;
3. whether the material appears to pander or is used to titillate, or whether the material appears to have been presented for its shock value.[47]

A discussion of cases that address each of these factors follows.

Explicitness or Graphic Nature of Material

Generally, the more explicit or graphic the description or depiction, the greater the likelihood that the material will be deemed patently offensive and therefore indecent. For example, the Commission imposed a forfeiture on a university radio station for airing a rap song that included a line depicting anal intercourse.[48] In that case, the Commission determined that the song described sexual activities in graphic terms that were patently offensive and therefore indecent. Since the song was broadcast in the mid-afternoon, there was a reasonable risk that children were in the audience, thus giving rise to the Commission's action.[49]

Broadcasts need not be as graphic as the song in the above case to give rise to the imposition of an FCC forfeiture. Broadcasts consisting of double entendres or innuendos may also be deemed indecent if the "sexual or excretory import is unmistakable." The FCC issued a notice of apparent liability and imposed a forfeiture on several stations for airing a song that included the following lines: "I whipped out my Whopper and whispered, Hey, Sweettart, how'd you like to Crunch on my Big Hunk for a Million Dollar Bar? Well, she immediately went down on my Tootsie Roll and you know, it was like pure Almond Joy."[50] The Commission determined that the material was indecent even though it used candy bar names to substitute for sexual activities. In one notice concerning the broadcast of the song, the Commission stated that "[w]hile the passages arguably consist of double entendre and indirect references, the language used in each passage was understandable and clearly capable of specific sexual meaning and, because of the context, the sexual import was inescapable."[51] The nature of the lyrics, coupled with the fact that the song aired between 6 a.m. and 10 a.m., gave rise to the imposition of a forfeiture.

Dwelling or Repetition of Potentially Offensive Material

Repetition of and persistent focus on a sexual or excretory activity could "exacerbate the potential offensiveness of broadcasts." For example, the FCC issued a notice of apparent liability and imposed a forfeiture on a radio station that broadcast an extensive discussion of flatulence and defecation by radio personality "Bubba, the Love Sponge."[52] Though the broadcast did not contain any expletives, the Commission found that the material dwelt on excretory activities and therefore was patently offensive.

While repetition can increase the likelihood that references to sexual or excretory activities are deemed indecent, where such references have been made in passing or are fleeting in nature, the Commission has found that the reference was not indecent even when profanity has been used.[53] For example, the Commission determined that the following phrase — "The hell I did, I drove mother-f[***]er, oh." — uttered by an announcer during a radio morning show, was not indecent.[54] The Commission declined to take action regarding the broadcast because it contained only a "fleeting and isolated utterance . . . within the context of live and spontaneous programming."[55] Certain fleeting references may, however, be found indecent where other factors contribute to the broadcast's patent offensiveness. For example, the Commission has imposed forfeitures on stations for airing jokes that refer to sexual activities with children.[56]

Pandering or Titillating Nature of Material

In determining whether broadcast material is indecent, the Commission also looks to the purpose for which the material is being presented. Indecency findings generally involve material that is presented in a pandering or titillating manner, or material that is presented for the shock value of its language. For example, the Commission deemed a radio call-in survey about oral sex to be indecent based in part on the fact that the material was presented in a pandering and titillating manner.[57]

Whether a broadcast is presented in a pandering or titillating manner depends on the context in which the potentially indecent material is presented. Explicit images or graphic language does not necessarily mean that the broadcast is being presented in a pandering or titillating manner. For example, the Commission declined to impose a forfeiture on a television station for airing portions of a high school sex education class that included the use of "sex organ models to demonstrate the use of various birth control devices."[58] In dismissing the complaint, the Commission held that, "[a]lthough the program dealt with sexual issues, the material presented was clinical or instructional in nature and not presented in a pandering, titillating, or vulgar manner."[59]

GOLDEN GLOBE AWARDS DECISION

As noted above, on March 18, 2004, the Federal Communications Commission overturned an earlier decision by the Commission's Enforcement Bureau regarding the broadcast of the word "f[***]ing" during the 2003 Golden Globe Awards. In the earlier decision, the Enforcement Bureau had found that the broadcast of the program including the utterance did not violate federal restrictions regarding the broadcast of obscene and indecent material.[60] The Bureau dismissed the complaints primarily because the language in question did not describe or depict sexual or excretory activities or organs.

In its March 18 *Memorandum Opinion and Order*, the full Commission concluded that the broadcast of the Golden Globe Awards did include material that violated prohibitions on the broadcast of indecent and profane material.[61] In reversing the Bureau, the Commission determined that the "phrase at issue is within the scope of our indecency definition because it does depict or describe sexual activities."[62] Although the Commission "recognize[d] NBC's argument that the 'F-Word' here was used 'as an intensifier,'" it nevertheless concluded that, "given the core meaning of the 'F-Word,' any use of that word or a variation, in any context, inherently has a sexual connotation, and therefore falls within the first prong of our indecency definition."[63]

Upon finding that the phrase in question fell within the first prong of the definition of "indecency," the Commission turned to the question of whether the broadcast was patently offensive under contemporary community standards for the broadcast medium. The Commission determined that the broadcast was patently offensive, noting that "[t]he 'F-Word' is one of the most vulgar, graphic and explicit descriptions of sexual activity in the English language," and that "[t]he use of the 'F-Word' here, on a nationally telecast awards ceremony, was shocking and gratuitous."[64] The Commission also rejected "prior Commission and staff action [that] have indicated that isolated or fleeting broadcasts of the

'F-Word' such as that here are not indecent or would not be acted upon," concluding "that any such interpretation is no longer good law."[65] The Commission further clarified its position, stating "that the mere fact that specific words or phrases are not sustained or repeated does not mandate a finding that material that is otherwise patently offensive to the broadcast medium is not indecent."[66]

In addition to the determination that the utterance of the word "f[***]ing" during the Golden Globe Awards was indecent, the Commission also found, as an independent ground for its decision, that use of the word was "profane" in violation of 18 U.S.C. 1464.[67] In making this determination, the Commission cited dictionary definitions of "profanity" as "'vulgar, irreverent, or coarse language,'"[68] and a Seventh Circuit opinion stating that "profanity" is "'construable as denoting certain of those personally reviling epithets naturally tending to provoke violent resentment or denoting language so grossly offensive to members of the public who actually hear it as to amount to a nuisance.'"[69] The Commission acknowledged that its limited case law regarding profane speech has focused on profanity in the context of blasphemy, but stated that it would no longer limit its definition of profane speech in such manner. Pursuant to its adoption of this new definition of "profane," the Commission stated that, depending on the context, the "'F-Word' and those words (or variants thereof) that are as highly offensive as the 'F-Word'" would be considered "profane" if broadcast between 6 a.m. and 10 p.m.[70] The Commission noted that other words would be considered on a case-by-case basis.

The Second Circuit's decision in *Fox Television Stations, Inc. v. FCC*, discussed below at pages 16-17, did not involve the Bono case, but it effectively overturned it.

SUPER BOWL HALFTIME SHOW DECISION

As noted above, on September 22, 2004, the FCC released a *Notice of Apparent Liability for Forfeiture* imposing a $550,000 forfeiture on several Viacom-owned CBS affiliates for the broadcast of the Super Bowl XXXVIII halftime show on February 1, 2004, in which a performer's breast was exposed.[71] The Commission determined that the show, which was aired at approximately 8:30 p.m. Eastern Standard Time, violated its restrictions on the broadcast of indecent material.

In its analysis, the Commission determined that since the broadcast included a performance that culminated in "on-camera partial nudity," and thus satisfied the first part of the indecency analysis, further scrutiny was warranted to determine whether the broadcast was "patently offensive as measured by contemporary community standards for the broadcast medium."[72] The Commission found that the performance in question was "both explicit and graphic," and rejected the licensees' contention that since the exposure was fleeting, lasting only 19/32 of a second, it should not be deemed indecent.[73] In determining whether the material in question was intended to "pander to, titillate and shock the viewing audience," the Commission noted that the performer's breast was exposed after another performer sang, "gonna have you naked by the end of this song."[74] The Commission found that the song lyrics, coupled with simulated sexual activities during the performance and the exposure of the breast, indicated that the purpose of the performance was to pander to, titillate and shock the audience, and the fact that the actual exposure of the breast was brief, as noted above, was not dispositive.[75]

The Commission ordered each Viacom-owned CBS affiliate to pay the statutory maximum forfeiture of $27,500 for the broadcast, for a total forfeiture of $550,000. The forfeiture was imposed on the Viacom-owned affiliates because of Viacom's participation in and planning of the Super Bowl halftime show with MTV networks, another Viacom subsidiary.[76]

Following the issuance of the *Notice of Apparent Liability for Forfeiture*, the affiliates are "afforded a reasonable period of time (usually 30 days from the date of the notice) to show, in writing, why a forfeiture penalty should not be imposed or should be reduced, or to pay the forfeiture."[77] CBS filed an opposition to the Notice of Apparent Liability on November 5, 2004. The opposition challenged the forfeiture on various grounds, including that the test for indecency was not met and that the forfeiture violates the First Amendment.

On March 15, 2006, the FCC issued a *Forfeiture Order* imposing a mandatory forfeiture in the amount of $550,000 on CBS for the airing of the 2004 Super Bowl halftime show.[78] CBS appealed to the U.S. Court of Appeals for the Third Circuit, which heard oral arguments in the case on September 11, 2007.[79]

OTHER RECENT ENFORCEMENT ACTIONS

In addition to the *Order* regarding the 2004 Super Bowl halftime show, the FCC issued several *Notices of Apparent Liability* for various television broadcasts occurring between February 2, 2002, and March 8, 2005.[80]

Of the six programs for which a forfeiture was proposed, two of the complaints were based on the use of "indecent" language, two were based on sexually explicit images, and two programs were cited for both language and sexual innuendo.[81] In determining whether a forfeiture was appropriate, the Commission applied the modified analysis first used in the *Golden Globe Awards Order* with respect to language that is deemed "indecent," and in the *Super Bowl Halftime Show Order* regarding sexually explicit imagery.[82]

In addition to the Commission's recent actions with respect to televised programming, the Commission had previously imposed forfeitures on a number of radio stations for broadcast indecency.[83] We now discuss two of its more recent high-profile actions related to radio programming. Each of these actions resulted in a consent decree between the Commission and the broadcaster.

Infinity Broadcasting

On October 2, 2003, the Commission issued a *Notice of Apparent Liability* to Infinity Broadcasting for airing portions of the "Opie and Anthony Show" during which the hosts conducted a contest entitled "Sex for Sam" which involved couples having sex in certain "risky" locations throughout New York City in an effort to win a trip.[84] The couples, accompanied by a station employee, were to have sex in as many of the designated locations as possible. They were assigned points based on the nature of the location and the activities in which they engaged. The station aired discussions between the hosts of the show and the station employee accompanying the couples which consisted of descriptions of the sexual activities of the participating couples and the locations in which they engaged in sexual activities. One discussion involved an description of a couple apparently engaging in sexual activities in St. Patrick's Cathedral.

The Commission determined that the broadcast made "graphic and explicit references to sexual and excretory organs and activity" despite the fact that colloquial terms, rather than explicit or graphic terms, were used in the descriptions. The Commission found that "[t]o the extent that the colloquial terms that the participants used to describe organs and activities could be described as innuendo rather than as direct references, they are nonetheless sufficient to render the material actionably indecent because the 'sexual [and] excretory import' of those references was 'unmistakable.'"[85] The Commission also found that the hosts of the show "dwelled at length on and referred repeatedly to sexual or excretory activities and organs," and that "the descriptions of sexual and excretory activity and organs were not in any way isolated and fleeting."[86]

On November 23, 2004, the FCC entered into a consent decree with Infinity regarding the Opie and Anthony *NAL*.[87] Pursuant to the decree, Infinity, a subsidiary of Viacom, agreed to make a voluntary contribution to the United States Treasury in the amount of $3.5 million and to adopt a company-wide compliance plan for the purpose of preventing the broadcast of indecent material. As part of the company-wide plan, Viacom agreed to install delay systems to edit "potentially problematic" live programming and to conduct training with respect to indecency regulations for all of its on-air talent and employees who participate in programming decisions.

Clear Channel Broadcasting

On January 27, 2004, the Commission issued a *Notice of Apparent Liability* to Clear Channel Broadcasting for repeated airings of the "Bubba, the Love Sponge" program which included indecent material.[88] The Commission found that all the broadcasts in question involved "conversations about such things as oral sex, penises, testicles, masturbation, intercourse, orgasms and breasts."[89] The Commission determined that each of the broadcasts in question contained "sufficiently graphic and explicit references," which were generally repeated throughout the broadcast in a pandering and titillating manner.

In one broadcast, the station aired a segment involving skits in which the voices of purported cartoon characters talk about drugs and sex.[90] The skits were inserted between advertisements for Cartoon Network's Friday-night cartoons. The Commission determined that "the use of cartoon characters in such a sexually explicit manner during hours of the day when children are likely to be listening is shocking and makes this segment patently offensive."[91] The Commission also cited the "calculated and callous nature of the stations' decision to impose this predictably offensive material upon young, vulnerable listeners" as "weighing heavily" in its determination.[92]

On April 8, 2004, the Commission released another *Notice of Apparent Liability* against Clear Channel Communications for airing allegedly indecent material during the "Howard Stern Show."[93] For the first time, the Commission sought to impose separate statutory maximum forfeitures for each indecent utterance during the program in question, rather than imposing a single fine for the entire program.[94]

The Commission entered into a consent decree with Clear Channel on June 9, 2004. The decree requires Clear Channel to make a "voluntary contribution" of $1.75 million to the United States Treasury and outlines "a company-wide compliance plan for the purpose of preventing the broadcast over radio or television of material violative of the indecency laws."[95] As part of the compliance plan, Clear Channel will "conduct training on obscenity and indecency for all on-air talent and employees who materially participate in programming decisions, which will include tutorials regarding material that the FCC does not permit broadcasters to air."[96] The plan also requires Clear Chanel to suspend any employee accused of airing, or who materially participates in the decision to air, obscene or indecent material while an investigation is conducted following the issuance of a *Notice of Apparent Liability*. Such employees will be terminated without delay if the *NAL* results in enforcement action by the FCC.

Fox Television Stations, Inc. v. FCC

On December 20, 2006, the U.S. Court of Appeals for the Second Circuit heard oral arguments in a case brought by four television networks — CBS, ABC, NBC, and Fox Television — challenging the FCC's authority to regulate "indecent" material. The FCC had taken action against, among other broadcasts, two award shows, described in an Associated Press article as, "a December 9, 2002, broadcast of the Billboard Music Awards in which singer Cher used the phrase, 'F--- 'em,' and a December 10, 2003, Billboard awards show in which reality show star Nicole Richie said: 'Have you ever tried to get cow s--- out of a Prada purse? It's not so f------ simple.'"[97] These incidents raise the same questions that FCC's action against the Bono expletive raised: whether a fleeting isolated expletive is "indecent" under federal law, and, if so, whether the First Amendment permits the FCC to enforce the law by punishing broadcasters for such utterances.

On June 4, 2007, the Second Circuit, in a 2-1 decision, found "that the FCC's new policy regarding 'fleeting expletives' represents a significant departure from positions previously taken by the agency and relied on by the broadcast industry. We further find that the FCC has failed to articulate a reasoned basis for this change in policy. Accordingly, we hold that the FCC's new policy regarding 'fleeting expletives' is arbitrary and capricious under the Administrative Procedure Act."[98]

The court rejected several reasons that the FCC gave for departing from its previous positions. The FCC, for example, attempted to justify its stance on fleeting expletives on the basis that "granting an automatic exemption for 'isolated or fleeting' expletives unfairly forces viewers (including children) to take 'the first blow.'"[99] The Second Circuit could not "accept this argument as a reasoned basis justifying the Commission's new rule. First, the Commission provides no reasonable explanation for why it has changed its perception that a fleeting expletive was not a harmful 'first blow' for the nearly thirty years between *Pacifica* and *Golden Globes*. More problematic, however, is that the 'first blow' theory bears no rational connection to the Commission's actual policy regarding fleeting expletives. As the FCC itself stressed during oral argument in this case, the Commission does not take the position that *any* occurrence of an expletive is indecent or profane under its rules. For example, although 'there is no outright news exemption from our indecency rules,' . . . the Commission will apparently excuse an expletive when it occurs during a '*bona fide* news interview . . . .'"[100]

Another FCC argument that the court rejected was that it is "difficult (if not impossible) to distinguish whether a word is being used as an expletive or as a literal description of sexual or excretory functions."[101] The court found, "This defies any commonsense understanding of these words, which, as the general public knows, are often used in everyday conversation without any 'sexual or excretory' meaning. Bono's exclamation . . . is a prime example of a non-literal use of the 'F-Word' that has no sexual connotation."[102]

The court also noted that the FCC's decision was "devoid of any evidence that suggests a fleeting expletive is harmful," which "would seen to be particularly relevant today when children likely hear this language far more often from other sources than they did in the 1970s when the Commission first began sanctioning indecent speech." In addition, the court found, "The Commission's new approach to profanity is supported by even less analysis, reasoned or not."[103]

Having overturned the FCC policy on statutory grounds, the court had no occasion to decide whether it also violated the First Amendment. It explained, however, why it was "skeptical that the Commission can provide a reasonable explanation for its 'fleeting expletive' regime that would pass constitutional muster."[104] The final section of this report examines this aspect of the court's opinion.

CONGRESSIONAL RESPONSE

109th Congress

Legislation addressing broadcast indecency was introduced during the 109th Congress.

H.R. 310, 109th Congress, entitled the Broadcast Decency Enforcement Act of 2005, would increase penalties for the broadcast of obscene, indecent, or profane material to a maximum of $500,000 for each violation, and would provide penalties for nonlicensees, such as artists and performers.[105] In addition to the increased forfeiture amount, the Federal Communications Commission would be able to require licensees to broadcast public service announcements that serve the educational and informational needs of children and require such announcements to reach an audience up to five times greater than the size of the audience that is estimated to have been reached by the obscene, indecent or profane broadcast. Also, any violation of the Commission's indecency regulations could be considered when determining whether to grant or renew a broadcast license, and three or more indecency violations would trigger a license revocation proceeding. The legislation would also require the Commission to act upon allegations of indecency within 180 days of the receipt of the allegation.

On February 9, 2005, the House Committee on Energy and Commerce reported H.R. 310 favorably without amendment. The House passed H.R. 310 on February 16, with one amendment. The amendment made several nonsubstantive technical changes and added a section requiring the FCC to revise its policy statement regarding broadcast indecency within nine months of the date of enactment of the legislation and at least once every three years thereafter[106]

H.R. 1420, 109th Congress, entitled the Families for ED Advertising Decency Act, would require the Federal Communications Commission to revise its interpretations of, and enforcement policies concerning, the broadcast of indecent material to treat as indecent any advertisement for a medication for the treatment of erectile dysfunction, thus prohibiting the airing of such advertisements between the hours of 6 a.m. and 10 p.m.

H.R. 1440, 109th Congress, entitled the Stamp Out Censorship Act of 2005, would prohibit the Federal Communications Commission from imposing forfeitures for indecency on nonbroadcast programming.

S. 193, 109th Congress, also entitled the Broadcast Decency Enforcement Act of 2005, was signed into law by the President on June 15, 2006, and became P.L. 109-235. It increases the maximum forfeiture for the broadcast of obscene, indecent, or profane material to $325,000 for each violation, with a cap of $3,000,000 for any single act or failure to act.[107]

S. 616, 109th Congress, entitled the Indecent and Gratuitous and Excessively Violent Programming Control Act of 2005, would, among other things, require the Federal Communications Commission to study the effectiveness of measures used by multichannel video programming distributors to protect children from exposure to indecent and violent programming. If the Commission were to determine that such measures were not effective, the legislation directs the Commission to initiate a rulemaking proceeding to adopt measures to protect children from indecent programming carried by multichannel video programming distributors during the hours when children are reasonably likely to be a substantial portion of the television audience. The bill would also increase the penalty for the broadcast of obscene, indecent, or profane language or images to $500,000, with a maximum forfeiture for any violations occurring in a given 24-hour period set at $3,000,000. The Commission would be required to take certain factors into consideration when imposing a forfeiture, and would be required to conduct public hearings or forums prior to the imposition of a forfeiture. Additionally, broadcast licensees would be allowed to preempt programming from a network organization that it deems obscene, and all television and radio broadcast licensees and multichannel video programming distributors would be required to provide a warning of the specific content of each recorded or scripted program it broadcasts.

110th Congress

Fewer bills have been introduced relating to broadcast indecency during the 110th Congress than during the 109th Congress.

H.R.2738, 110th Congress, entitled the Family and Consumer Choice Act of 2007, would not apply to broadcast indecency but would regulate indecency by requiring multichannel video programming distributors (cable operators, broadcast satellite service providers, and multichannel multipoint distribution services, among others) to choose among three options. First, providers could choose not to transmit any material defined as indecent under the FCC's broadcast indecency policies on any channel on their expanded basic tier between certain hours; second, providers could choose to scramble any channel, free of charge, that a subscriber does not wish to receive; or, third, providers could make available a "family tier" of channels that would consist only of channels that do not transmit indecent programming. This bill was referred to the House Committee on Energy and Commerce on June 15, 2007.

S.602, 110th Congress, entitled the Child Safe Viewing Act of 2007, would encourage the use of advanced blocking technologies. The bill would require the FCC to examine blocking technologies that could be utilized across various platforms, including wired, wireless, and Internet, and to consider measures that would encourage or require the use of these technologies. Advanced blocking technologies aid parents in denying children access to indecent or objectionable programming available over broadcast television, cable, Internet, and other media.

S. 602 was introduced on February 15, 2007, and referred to the Senate Committee on Commerce, Science, and Transportation.

S.1780, 110th Congress, entitled the Protecting Children from Indecent Programming Act, would amend section 16 of the Public Telecommunications Act of 1992, 47 U.S.C. § 303 note, to require the FCC to "maintain a policy that a single word or image may constitute indecent programming."

On July 12, 2007, the bill was referred to the Senate Committee on Commerce, Science and Transportation. The bill was reported to the Senate without amendment after markup on July 19, 2007.

WOULD PROHIBITING THE BROADCAST OF "INDECENT" WORDS REGARDLESS OF CONTEXT VIOLATE THE FIRST AMENDMENT?

In 1978, in *Federal Communications Commission v. Pacifica Foundation*, the Supreme Court upheld, against a First Amendment challenge, an action that the FCC took against a radio station for broadcasting a recording of George Carlin's "Filthy Words" monologue at 2 p.m.[108] The Court has not decided a case on the issue of "indecent" speech on broadcast radio or television since then, but it did cite *Pacifica* with approval in 1997, when, in *Reno v. ACLU*, it contrasted regulation of the broadcast media with regulation of the Internet.[109] Nevertheless, the Court in *Reno* did not *hold* that *Pacifica* remains good law, and arguments have been made that the proliferation of cable television channels has rendered archaic *Pacifica*'s denial of full First Amendment rights to broadcast media.

Even if *Pacifica* remains valid in this respect, *Pacifica* did not hold that the First Amendment permits the ban either of an occasional expletive on broadcast media, or of programs that would not be likely to attract youthful audiences, even if such programs contain "indecent" language. On these points, Justice Stevens wrote for the Court in *Pacifica*:

> It is appropriate, in conclusion, to emphasize the narrowness of our holding. This case does not involve a two-way radio conversation between a cab driver and a dispatcher, or a telecast of an Elizabethan comedy. We have not decided that an occasional expletive in either setting would justify any sanction. . . . The time of day was emphasized by the Commission. The content of the program in which the language is used will also affect the composition of the audience. . . .[110]

In a footnote to the last sentence of this quotation, the Court added: "Even a prime-time recitation of Geoffrey Chaucer's Miller's Tale would not be likely to command the attention of many children. . . ."[111] At the same time, Justice Stevens acknowledged that the Carlin monologue has political content: "The monologue does present a point of view; it attempts to show that the words it uses are 'harmless' and that our attitudes toward them are 'essentially silly.' The Commission objects, [however,] not to this point of view, but to the way in which it is expressed."[112] The Court commented: "If there were any reason to believe that the Commission's characterization of the Carlin monologue as offensive could be traced to its political content — or even to the fact that it satirized contemporary attitudes about four-letter words — First Amendment protection might be required."[113]

There appears to be some tension between this comment and the Court's remark about Chaucer, as any attempt to censor Chaucer would presumably also be based not on its ideas but on the way its ideas are expressed. But, as noted above, the Court's remark about Chaucer was a footnote to its comment that "[t]he content of the program in which the language is used will also affect the composition of the audience. . . ." Therefore, the difference that Justice Stevens apparently perceived between Chaucer and Carlin was that, even if both have literary, artistic, or political value, only the latter would be likely to attract a youthful audience. Arguably, then, *Pacifica* would permit the censorship, during certain hours, of the broadcast even of works of art that are likely to attract a youthful audience.[114]

If so, this would be contrary to the Court's opposition, in other contexts, to the censorship of works of art. The Court has held that even "materials [that] depict or describe patently offensive 'hard core' sexual conduct," which would otherwise be obscene, may not be prohibited if they have "serious literary, artistic, political, or scientific value."[115] In addition, the "harmful to minors" statutes of the sort that the Supreme Court upheld in *Ginsberg v. New York* generally define "harmful to minors" to parallel the Supreme Court's definition of "obscenity," and thus prohibit distributing to minors only material that lacks serious value for minors.[116] This suggests that, if the FCC or Congress prohibited the broadcast during certain hours of "indecent" words regardless of context, the Court might be troubled by the prohibition's application to works with serious value, even though *Pacifica* allowed the censorship of Carlin's monologue, despite its apparently having serious value.

Yet, as noted, Justice Stevens' expressed a distinction in *Pacifica* between a point of view and the way in which it is expressed, and, though a majority of the justices did not join the part of the opinion that drew this distinction, a majority of the justices, by concurring in *Pacifica*'s holding, indicated that the political (or literary or artistic) content of Carlin's monologue did not prevent its censorship during certain hours on broadcast radio and television. Therefore, it appears that, in deciding the constitutionality of an FCC or a congressional action prohibiting the broadcasting, during certain hours, of material with "indecent" words, the Court might be troubled by its application to works with serious value only if those works would, like Chaucer's, not likely attract a substantial youthful audience.

In sum, the Court did not hold that the FCC could prohibit an occasional expletive, and did not hold that the FCC could prohibit offensive words in programs — even prime-time programs — that children would be unlikely to watch or listen to. The Court did not hold that the FCC could *not* take these actions, as the question whether it could was not before the Court. But the Court's language quoted above renders *Pacifica* of uncertain precedential value in deciding whether a ban, during certain hours, on the broadcast of "indecent" words regardless of context would be constitutional.

In the "Filthy Words" monologue, as the Supreme Court described it, George Carlin "began by referring to his thoughts about 'the words you couldn't say on the public, ah, airwaves, um, the ones you definitely wouldn't say, ever.' He proceeded to list those words and repeat them over and over in a variety of colloquialisms." The FCC, at the time, used essentially the same standard for "indecent" that it uses today: "[T]he concept of 'indecent' is intimately connected with the exposure of children to language that describes, in terms patently offensive as measured by contemporary community standards for the broadcast medium, sexual or excretory activities and organs. . . ."[117]

Most of Carlin's uses of the "filthy words," it appears from reading his monologue, which is included as an appendix to the Court's opinion, seem designed to show the words' multiple uses, apart from describing sexual or excretory activities or organs. Nevertheless, "the Commission concluded that certain words depicted sexual or excretory activities in a patently offensive manner. . . ."[118] Therefore, one might argue that, even if, under *Pacifica*, the First Amendment does not protect, during certain hours, the use on broadcast media of words that depict sexual or excretory activities in a patently offensive manner, it nevertheless might protect the use of those same words "as an adjective or expletive to emphasize an exclamation" (to quote the FCC Enforcement Bureau's opinion in the Bono case).

A counterargument might be that, in *Pacifica*, the Court noted that "the normal definition of 'indecent' merely refers to nonconformance with accepted standards of morality."[119] This suggests the possibility that the Court would have ruled the same way in *Pacifica* if the FCC had defined "indecent" loosely enough to include the use of a patently offensive word "as an adjective or expletive to emphasize an exclamation." But this is speculative, as the Court did not so rule. Further, as noted above, Court emphasized the narrowness of its holding, noting that it had "not decided that an occasional expletive . . . would justify any sanction. . . ."

On what basis did the Court in *Pacifica* find that the FCC's action did not violate the First Amendment? In Part IV-C of opinion, which was joined by a majority of the justices, Justice Stevens wrote:

[O]f all forms of communication, it is broadcasting that has received the most limited First Amendment protection. Thus, although other speakers cannot be licensed except under laws that carefully define and narrow official discretion, a broadcaster may be deprived of his license and his forum if the Commission decides that such an action would serve "the public interest, convenience, and necessity." Similarly, although the First Amendment protects newspaper publishers from being required to print the replies of those whom they criticize,

Miami Herald Publishing Co. v. Tornillo, 418 U.S. 241, it affords no such protection to broadcasters; on the contrary, they must give free time to the victims of their criticism. *Red Lion Broadcasting Co. v. FCC*, 395 U.S. 367.

The reasons for these distinctions are complex, but two have relevance to the present case. First, the broadcast media have established a uniquely pervasive presence in the lives of all Americans. Patently offensive, indecent material presented over the airwaves confronts the citizen, not only in public, but in the privacy of the home, where the individual's right to be left alone plainly outweighs the First Amendment rights of an intruder. *Rowan v. Post Office Dept.*, 397 U.S. 728. . . . To say that one may avoid further offense by turning off the radio when he hears indecent language is like saying that the remedy for an assault is to run away after the first blow.

Second, broadcasting is uniquely accessible to children, even those too young to read. . . . Bookstores and motion picture theaters . . . may be prohibited from making indecent material available to children. We held in *Ginsberg v. New York*, 390 U.S. 629, that the government's interest in the "well-being of its youth" and in supporting "parents' claim to authority in their own household" justified the regulation of otherwise protected expression. . . .[120]

In sum, the Court held that, on broadcast radio and television, during certain times of day, certain material may be prohibited because (1) it is patently offensive and indecent, and (2) it threatens the well-being of minors and their parents' authority in their own household. This raises the question of the extent to which the Court continues to allow the government (1) to treat broadcast media differently from other media, and (2) to censor speech on the ground that it is patently offensive and indecent, or threatens the well-being of minors and their parents' authority in their own household.

Broadcast Media

In *Red Lion Broadcasting Co. v. FCC*, which the Court cited in the above quotation from *Pacifica*, the Court upheld the FCC's "fairness doctrine," which "imposed on radio and television broadcasters the requirement that discussion of public issues be presented on broadcast stations, and that each side of those issues must be given fair coverage."[121] The reason that the Court upheld the imposition of the fairness doctrine on broadcast media, though it would not uphold its imposition on print media, is that "[w]here there are substantially more individuals who want to broadcast than there are frequencies to allocate, it is idle to posit an unabridgeable First Amendment right to broadcast comparable to the right of every individual to speak, write, or publish."[122] "Licenses to broadcast," the Court added, "do not confer ownership of designated frequencies, but only the temporary privilege of using them. 47 U.S.C. § 301. Unless renewed, they expire within three years. 47 U.S.C. § 307(d). The statute mandates the issuance of licenses if the 'public convenience, interest, or necessity will be served thereby.' 47 U.S.C. § 307(a)."[123]

The Court in *Red Lion* then noted:

> It is argued that even if at one time the lack of available frequencies for all who wished to use them justified the Government's choice of those who would best serve the public interest . . . this condition no longer prevails so that continuing control is not justified. To this there are several answers. Scarcity is not entirely a thing of the past.[124]

With the plethora of cable channels today, has spectrum scarcity now become a thing of the past? In *Turner Broadcasting System, Inc. v. FCC*, the Court held that the scarcity rationale does not apply to cable television:

> [C]able television does not suffer from the inherent limitations that characterize the broadcast medium . . . [S]oon there may be no practical limitation on the number of speakers who may use the cable medium. Nor is there any danger of physical interference between two cable speakers attempting to use the same channel. In light of these fundamental technological differences between broadcast and cable transmission, application of a more relaxed standard of scrutiny adopted in *Red Lion* and the other broadcast cases is inapt when determining the First Amendment validity of cable regulation."[125]

One might argue that, if the scarcity rationale does not apply to cable television, then it should not apply to broadcast television either, because a person who because of scarcity cannot start a broadcast channel can start a cable channel.[126] But the Court has not ruled on the question; in *Turner* it wrote: "Although courts and commentators have criticized the scarcity rationale since its inception, we have declined to question its continuing validity as support for our broadcast jurisprudence, and see no reason to do so here."[127]

In 1987, however, the FCC abolished the fairness doctrine, on First Amendment grounds, noting that technological developments and advancements in the telecommunications marketplace have provided a basis for the Supreme Court to reconsider its holding in *Red Lion*. The FCC's decision was upheld by the U.S. Court of Appeals for the District of Columbia, and the Supreme Court declined to review the case.[128] The court of appeals did not rule on constitutional grounds, but rather concluded "that the FCC's decision that the fairness doctrine no longer served the public interest was neither arbitrary, capricious nor an abuse of discretion, and [we] are convinced that it would have acted on that finding to terminate the doctrine even in the absence of its belief that the doctrine was no longer constitutional."[129]

But, whether or not spectrum scarcity has become a thing of the past, it would apparently would not today justify governmental restrictions on "indecent" speech. This is because, subsequent to the Court in *Turner* declining to question the applicability of the scarcity rationale to broadcast media, a plurality of justices noted, in *Denver Area Educational Telecommunications Consortium, Inc. v. FCC*, that, though spectrum scarcity continued to justify the "structural regulations at issue there [in *Turner*] (the 'must carry' rules), it has little to do with a case that involves the effects of television viewing on children. Those effects are the result of how parents and children view television programming, and how pervasive and intrusive that programming is. In that respect, cable and broadcast television differ little, if at all."[130] The plurality therefore upheld a federal statute that permits cable operators to prohibit indecent material on leased access channels. Thus, it appears that the Court today would not cite spectrum scarcity to justify restrictions on "indecent" material on broadcast media, but it might cite broadcast media's pervasiveness and intrusiveness.

Subsequent to *Denver Area*, in *United States v. Playboy Entertainment Group, Inc.*, the Court held that cable television has full First Amendment protection; i.e., content-based restrictions on cable television receive strict scrutiny.[131] Thus, if, as the Court said in *Denver Area*, cable and broadcast media differ little, if at all, with respect to the regulation of "indecent" material, and, if, as the Court said in *Playboy*, cable television receives strict scrutiny, then, arguably, broadcast media would also receive strict scrutiny with regard to restrictions on "indecent" material.[132] It is possible, however, that, if cable and broadcast media differ little, then the Court might apply *Pacifica* to both broadcast and cable, rather than to neither.[133] In any event, as noted above, even if the Court were to continue to apply *Pacifica* to restrictions on broadcast media, this does not necessarily mean that it would uphold a ban on the broadcast of "indecent" language regardless of context, as *Pacifica* did not hold that an occasional expletive would justify a sanction.

Strict Scrutiny

We now consider the analysis that the Court might apply if it chooses not to apply *Pacifica* in deciding the constitutionality of a ban on the broadcast of "indecent" language regardless of context. The Court in *Pacifica*, as noted, offered two reasons why the FCC could prohibit offensive speech on broadcast media: "First, the broadcast media have established a uniquely pervasive presence in the lives of all Americans. Patently offensive, indecent material presented over the airwaves confronts the citizen, not only in public, but in the privacy of the home. . . . Second, broadcasting is uniquely accessible to children, even those too young to read," and the government has an interest in the "well-being of its youth"and "in supporting 'parents' claim to authority in their own household.'" The first of these reasons apparently refers to adults as well as to children.

Ordinarily, when the government restricts speech, including "indecent" speech, on the basis of its content, the restriction, if challenged, will be found constitutional only if it satisfies "strict scrutiny."[134] This means that the government must prove that the restriction serves "to promote a compelling interest" and is "the least restrictive means to further the articulated interest."[135] The Court in *Pacifica* did not apply this test or any weaker First Amendment test, and did not explain why it did not. Its reason presumably was that the FCC's action restricted speech only on broadcast media. If, however, the Court were not to apply *Pacifica* in determining the constitutionality of a ban, during certain hours, on the broadcast of "indecent" language regardless of context, then it would apparently apply strict scrutiny.

If the Court were to apply strict scrutiny in making this determination, it seems unlikely that it would find the first reason cited in *Pacifica* — sparing citizens, including adults, from patently offensive or indecent words — to constitute a compelling governmental interest. The Court has held that the government may not prohibit the use of offensive words unless they "fall within [a] relatively few categories of instances," such as obscenity, fighting words, or words "thrust upon unwilling or unsuspecting viewers."[136]

If the Court were to apply strict scrutiny in determining the constitutionality of a ban, during certain hours, on the broadcast of "indecent" language regardless of context, it also might not find the second reason cited in *Pacifica* — protecting minors from patently offensive and indecent words and "supporting 'parents' claim to authority in their own household'" — to constitute a compelling governmental interest. When the Court considers the constitutionality of a restriction on speech, it ordinarily — even when the speech lacks full First Amendment protection and the court applies less than strict scrutiny — requires the government to "demonstrate that the recited harms are real, not merely conjectural, and that the regulation will in fact alleviate these harms in a direct and material way."[137] With respect to restrictions designed to deny minors access to sexually explicit material, by contrast, the courts appear to assume, without requiring evidence, that such material is harmful to minors, or to consider it "obscene as to minors," even if it is not obscene as to adults, and therefore not entitled to First Amendment protection with respect to minors, whether it is harmful to them or not.[138] A word used as a mere adjective or expletive, however, arguably does not constitute sexually oriented material.[139] Therefore, if a court applied strict scrutiny to decide the constitutionality of a ban, during certain hours, on the broadcast of "indecent" words regardless of context, then, in determining the presence of a compelling interest, the court might require the government to "demonstrate that the recited harms are real, not merely conjectural, and that the regulation will in fact alleviate these harms in a direct and material way." This could raise the question, not raised in *Pacifica*, of whether hearing such words is harmful to minors. More precisely, it might raise the question of whether hearing such words on broadcast radio and television is harmful to minors, even in light of the opportunities for minors to hear such words elsewhere. If the government failed to prove that hearing certain words on broadcast radio or television is harmful to minors, then a court would not find a compelling interest in censoring those words and might strike down the law.

It might still uphold the law, however, if it found that the law served the government's interest "in supporting 'parents' claim to authority in their own household,'" and that this is a compelling interest independent from the interest in protecting the well-being of minors. In *Ginsberg v. New York*, the Court referred to the state's interest in the well-being of its youth as "independent" from its interest in supporting "parents' claim to authority in their own household to direct the rearing of their children."[140] The holding in *Ginsberg*, however, did not turn on whether these interests are independent, and one might argue that they are not, because the government's interest in supporting parents lies in assisting them in protecting their children from harmful influences. If "indecent" words are not a harmful influence, then, arguably, the government has no interest, sufficient to override the First Amendment, in supporting parents in their efforts to prevent their children's access to them. It has also been argued that "a law that effectively *bans* all indecent programming . . . does not facilitate parental supervision. In my view, my right as a parent has been preempted, not facilitated, if I am told that certain programming will be banned from my . . . television. Congress cannot take away my right to decide what my children watch, absent some showing that my children are in fact at risk of harm from exposure to indecent programming."[141]

If the government could persuade a court that a ban, during certain hours, on the broadcast of "indecent" words regardless of context serves a compelling interest —either in protecting the well-being of minors or in supporting parents' claim to authority — the government would then have to prove that the ban was the least restrictive means to advance that interest. This might raise questions such as whether it is necessary to prohibit particular words on weekdays during school hours, solely to protect pre-school children and children who are home sick some days. In response to this question, the government could note that the broadcast in *Pacifica* was at 2 p.m. on a Tuesday, but was nevertheless considered a "time[ ] of the day when there is a reasonable risk that children may be in the audience."[142] More significantly, however, a court might find a ban too restrictive because it would prohibit the broadcast, between certain hours, of material, including works of art and other material with serious value, that would not attract substantial numbers of youthful viewers or listeners.

In conclusion, it appears that, if a court were to apply strict scrutiny to determine the constitutionality of a ban on the broadcast of "indecent" language regardless of context, then it might require the government to "demonstrate that the recited harms are real, not merely conjectural, and that the regulation will in fact alleviate these harms in a direct and material way." This would mean that the government would have to demonstrate a compelling governmental interest, such as that hearing "indecent" words on broadcast radio and television is harmful to minors, despite the likelihood that minors hear such words elsewhere, or that banning "indecent" words is necessary to support parents' authority in their own household. If the government could not demonstrate a compelling governmental interest, then the court might find the ban unconstitutional. Even if the government could demonstrate a compelling interest, a court might find the ban unconstitutional if it applied to material with serious value, at least if such material would not attract substantial numbers of youthful viewers or listeners.

Whether a court would apply strict scrutiny would depend upon whether, in light of the proliferation of cable television, it finds *Pacifica* to remain applicable to broadcast media. If a court does find that *Pacifica* remains applicable to broadcast media, then the court would be faced with questions that *Pacifica* did not decide: whether, on broadcast radio and television during hours when children are likely to be in the audience, the government may prohibit an "indecent" word used as an occasional expletive, or in material that would not attract substantial numbers of youthful viewers or listeners.

Second Circuit's Dicta in Fox Television Stations, Inc. V. FCC

As noted at pages 16-18 of this report, on June 4, 2007, the Second Circuit, in a 2-1 decision, found "that the FCC's new policy regarding 'fleeting expletives' represents a significant departure from positions previously taken by the agency and relied on by the broadcast industry. We further find that the FCC has failed to articulate a reasoned basis for this change in policy. Accordingly, we hold that the FCC's new policy regarding 'fleeting expletives' is arbitrary and capricious under the Administrative Procedure Act."[143] As also noted above, the Second Circuit, having overturned the FCC policy on statutory grounds, had no occasion to decide whether it also violated the First Amendment. In dicta, however, it explained why it was "skeptical that the Commission can provide a reasonable explanation for its 'fleeting expletive' regime that would pass constitutional muster.'"

The court wrote that it was

> sympathetic to the Networks' contention that the FCC's indecency test is undefined, indiscernible, inconsistent, and consequently, unconstitutionally vague. . . . We also note that the FCC's indecency test raises the separate constitutional question of whether it permits the FCC to sanction speech based on its subjective view of the merit of that speech. It appears that under the FCC's current indecency regime, any and all uses of an expletive is presumptively indecent and profane with the broadcaster then having to demonstrate to the satisfaction of the Commission, under an unidentified burden of proof, that the expletives were "integral" to the work. In the licensing context, the Supreme Court has cautioned against speech regulations that give too much discretion to government officials. . . . Finally, we recognize that there is some tension in the law regarding the appropriate level of First Amendment scrutiny. In general, restrictions on First Amendment liberties prompt courts to apply strict scrutiny. . . . At the same time, however, the Supreme Court has also considered broadcast media exceptional. . . . Nevertheless, we would be remiss not to observe that it is increasingly difficult to describe the broadcast media as uniquely pervasive and uniquely accessible to children, and at some point in the future, strict scrutiny may properly apply in the context of regulating broadcast television.[144]

REFERENCES

[1]   The FCC's indecency regulations only apply to broadcast radio and television, and not to satellite radio or cable television. The distinction between broadcast and cable television arises in part from the fact that the rationale for regulation of broadcast media — the dual problems of spectrum scarcity and signal interference — do not apply in the context of cable. As a result, regulation of cable television is entitled to heightened First Amendment scrutiny. *See Turner Broadcasting v. Federal Communications Commission*, 512 U.S. 622 (1994). Cable television is also distinguished from broadcast television by the fact that cable involves a voluntary act whereby a subscriber affirmatively chooses to bring the material into his or home. *See Cruz v. Ferre*, 755 F.2d 1415 (11th Cir. 1985).

[2]   See CRS Report RL33170, *Constitutionality of Applying the FCC's Indecency Restriction to Cable Television*, by Henry Cohen.

[3]   See *In the Matter of Complaints Against Various Broadcast Licensees Regarding Their Airing of the "Golden Globe Awards" Program*, 18 F.C.C. Rcd. 19859 (2003).

[4]   *Id.* at 2.

[5]   *Id.*

[6]   *Id.* at 3.

[7]   *Id.*

[8]   "FCC Chairman Seeks Reversal on Profanity," *Washington Post*, January 14, 2004, at E01.

[9]   *In the Matter of Complaints Against Various Broadcast Licensees Regarding Their Airing of the "Golden Globe Awards" Program*, File No. EB-03-IH-0110 (March 18, 2004).

[10]   [http://hraunfoss.fcc.gov/edocs_public/attachmatch/DOC-243435A1.pdf].

[11]   *Complaints Against Various Television Licensees Concerning Their February 1, 2004, Broadcast of the Super Bowl XXXVIII Halftime Show*, File No. EB-04-IH-0011 (September 22, 2004) [http://www.fcc.gov/eb/Orders/2004/FCC-04-209A1.html].

[12]   *Id.*

[13]   18 U.S.C. § 1464. "Radio communication" includes broadcast television, as the term is defined as "the transmission by radio of writing, signs, signals, pictures, and sounds of all kinds." 47 U.S.C. § 153(33).

[14]   47 U.S.C. § 503(b).

[15]   438 U.S. 726 (1978).

[16]   The United State Court of Appeals for the District of Columbia Circuit had reversed the Commission's order. *See* 556 F.2d 9 (D.C. Cir. 1977). The Commission appealed that decision to the Supreme Court, which reversed the lower court's decision.

[17]   438 U.S. at 732.

[18]   *Id.*

[19]   *Id.* at 742.

[20]   *Id.* at 731; see, *In the Matter of a Citizen's Complaint Against Pacifica Foundation Station WBAI (FM), New York, New York*, 56 F.C.C.2d 94 (1975).

[21]   *In the Matter of Pacifica Foundation, Inc. d/b/a Pacifica Radio Los Angeles, California*, 2 F.C.C. Rcd. 2698 (1987). Two other orders handed down the same day

articulate the Commission's clarified indecency standard. *See also In the Matter of the Regents of the University of California*, 2 F.C.C. Rcd. 2703 (1987); *In the Matter of Infinity Broadcasting Corporation of Pennsylvania*, 2 F.C.C. Rcd. 2705 (1987).

[22] The Commission noted Arbitron ratings indicating that a number of children remain in the local audience well after 10 p.m. *See* 2 F.C.C. Rcd. 1698, 16.

[23] 2 F.C.C. Rcd. 2698, 12 and 15.

[24] 852 F.2d 1332, 1344 (1988).

[25] P.L. 100-459, § 608.

[26] *Enforcement of Prohibitions Against Broadcast Obscenity and Indecency*, 4 F.C.C. Rcd. 457 (1988).

[27] *Action for Children's Television v. Federal Communications Commission (ACT II)*, 932 F.2d 1504 (1991), *cert. denied*, 503 U.S. 913 (1992).

[28] *Id.* at 1509.

[29] P.L. 102-356, § 16, 47 U.S.C. § 303 note.

[30] *In the Matter of Enforcement of Prohibitions Against Broadcast Indecency in 18 U.S.C. 1464*, 8 F.C.C. Rcd. 704 (1993).

[31] *Action for Children's Television v. Federal Communications Commission*, 11 F.3d 170 (D.C. Cir. 1993).

[32] 8 F.C.C. Rcd. at 705-706.

[33] 11 F.3d at 171.

[34] *Id.*

[35] *Action for Children's Television v. Federal Communications Commission*, 15 F.3d 186 (D.C. Cir. 1994).

[36] *Action for Children's Television v. Federal Communications Commission (ACT III)*, 58 F.3d 654 (D.C. Cir. 1995), *cert. denied*, 516 U.S. 1043 (1996).

[37] 58 F.3d at 656.

[38] *Enforcement of Prohibitions Against Broadcast Indecency in 18 U.S.C. § 1464*, 10 F.C.C. Rcd. 10558 (1995); 47 C.F.R. 73.3999(b). Subsection (b) prohibits the broadcast of material which is obscene without any reference to time of day. Broadcast obscenity will not be discussed in this report. For more information on obscenity, see CRS Report 95-804, *Obscenity and Indecency: Constitutional Principles and Federal Statutes*, by Henry Cohen, and CRS Report 98-670, *Obscenity, Child Pornography, and Indecency: Recent Developments and Pending Issues*, by Henry Cohen.

[39] 60 FR 44439 (August 28, 1995).

[40] Enforcement actions based on televised broadcast indecency are rare. However, the Commission recently issued a *Notice of Apparent Liability* for the broadcast of indecent material during a televised morning news program. During the program, the show's hosts interviewed performers with a production entitled "Puppetry of the Penis," who appeared wearing capes but were otherwise nude. A performer's penis was exposed during the broadcast. *See In the Matter of Young Broadcasting of San Francisco, Inc.*, File No. EB-02-IH-0786 (January 27, 2004). *See also In the Matter of Complaints Against Various Licensees Regarding Their Broadcast of the Fox Television Network Program "Married by America" on April 7, 2003*, File No. EB-03-IH-0162 (October 12, 2004).

[41] Under 47 U.S.C. § 503(b)(2)(A), the maximum fine per violation is $25,000. However, the maximum forfeiture amount was increased to $32,500 pursuant to the Debt

Collection Improvement Act of 1996, Public Law 104-134, which amended the Federal Civil Monetary Penalty Inflation Adjustment Act of 1990, Public Law 101-410. *See* 47 C.F.R. § 1.80.

[42] Regulations set a maximum forfeiture of $325,000 for any single act or failure to act, which arguably limits the forfeiture for a single broadcast. *See* 47 C.F.R. § 1.80.

[43] 47 U.S.C. § 503(b)(1)(D) provides that the FCC may impose a forfeiture penalty upon any "person" who violates 18 U.S.C. § 1464, which makes it a crime to "utter" indecent language. In addition, 47 U.S.C. § 503(b)(6)(B) provides that the FCC may not impose a forfeiture penalty on a person who does not hold a broadcast station license if the violation occurred more than one year prior to the date of issuance of the required notice or notice of apparent liability. This suggests that the FCC may fine a performer if the violation occurred within one year of such date.

[44] *Complaints Against Various Television Licensees Concerning Their February 1, 2004, Broadcast of the Super Bowl XXXVIII Halftime Show*, File No. EB-04-IH-0011 (September 22, 2004) [http://www.fcc.gov/eb/Orders/2004/FCC-04-209A1.html].

[45] *See* note 41, *supra*.

[46] *See In the Matter of Industry Guidance on the Commission's Case Law Interpreting 18 U.S.C. § 1464 and Enforcement Policies Regarding Broadcast Indecency*, File No. EB-00-IH-0089 (April 6, 2001) [http://www.fcc.go v/eb/Orders/2001/fcc01090.html].

[47] *Id.*

[48] *Notice of Apparent Liability, State University of New York*, 8 F.C.C. Rcd. 456 (1993).

[49] *Id.*

[50] *Notice of Apparent Liability, KGB Incorporated*, 7 F.C.C. Rcd. 3207 (1992). *See also Great American Television and Radio Company, Inc.*, 6 F.C.C. Rcd. 3692 (1990); *WIOD, Inc.*, 6 F.C.C. Rcd. 3704 (1989).

[51] 6 F.C.C. Rcd. 3692.

[52] *Notice of Apparent Liability, Citicasters Co.*, 13 F.C.C. Rcd. 22004 (1998).

[53] The Commission has recently indicated that "the mere fact that specific words or phrases are not sustained or repeated does not mandate a finding that material that is otherwise patently offensive to the broadcast medium is not indecent." *In the Matter of Complaints Against Various Broadcast Licensees Regarding the Airing of the "Golden Globe Awards" Program*, File No. EB-03-IH-0110 (March 18, 2004). *See* section entitled "Golden Globe Awards Decision," below.

[54] *L.M. Communications of South Carolina, Inc.*, 7 F.C.C. Rcd. 1595 (1992).

[55] *Id.*

[56] *See Notice of Apparent Liability, Temple Radio, Inc.*, 12 F.C.C. Rcd. 21828 (1997); *Notice of Apparent Liability, EZ New Orleans, Inc.*, 12 F.C.C. Rcd. 4147 (1997).

[57] *Notice of Apparent Liability, Rusk Corporation, Radio Station KLOL*, 5 F.C.C. Rcd. 6332 (1990).

[58] *In the Matter of Application for Review of the Dismissal of an Indecency Complaint Against King Broadcasting Co.*, 5 F.C.C. Rcd. 2971 (1990).

[59] *Id.*

[60] *Id.*

[61] The Commission declined to impose a forfeiture on the broadcast licensees named in the complaint because they were not "on notice" regarding the new interpretations of the Commission's regulations regarding broadcast indecency and the newly adopted

definition of profanity. The Commission also indicated that it will not use its decision in this case adversely against the licensees during the license renewal process.

[62] *In the Matter of Complaints Against Various Broadcast Licensees Regarding Their Airing of the "Golden Globe Awards" Program*, File No. EB-03-IH-0110 at 4 (March 18, 2004).

[63] *Id*. Similarly, in March, 2006, the FCC decided that "s[***]" has an "inherently excretory connotation" and therefore could not be used from 6 a.m. to 10 p.m. See, *@$# and *% Ken Burns! PBS Scrubbing G.I. Mouths With Soap*, New York Observer, October 2, 2006, p. 1.

[64] *Id*. at 5.

[65] *Id*. at 6. *See* section entitled "Dwelling or Repetition of Potentially Indecent Material," above.

[66] *Id*.

[67] *Id*. at 7. It should be noted that, although in this case the Commission found that the broadcast in question was both indecent and profane, there are certain to be words that could be deemed "profane," but do not fit the Commission's definition of "indecent." Under the newly adopted definition of "profanity," many words could arguably be found "profane" because they provoke "violent resentment" or are otherwise "grossly offensive," but not be found "indecent" because they do not refer to any sexual or excretory activity or organ or even "inherently" have a sexual connotation, as the Commission found the phrase that Bono uttered to have. Presumably, it is these words that the Commission will consider on a case-by-case basis.

[68] *Id*. at 7, citing Black's Law Dictionary 1210 (6th ed. 1990) and American Heritage College Dictionary 1112 (4th ed. 2002).

[69] *Id*., citing *Tallman v. United States*, 465 F.2d 282, 286 (7th Cir. 1972).

[70] *Id*.

[71] *Complaints Against Various Television Licensees Concerning Their February 1, 2004, Broadcast of the Super Bowl XXXVIII Halftime Show*, File No. EB-04-IH-0011 (September 22, 2004) [http://www.fcc.gov/eb/Orders/2004/FCC-04-209A1.html].

[72] *Id*. at 11.

[73] *Id*. at 13.

[74] *Id*. at 14.

[75] *Id*.

[76] *Id*. at 17 - 24.

[77] 47 C.F.R. § 1.80(f)(3).

[78] *In the Matter of Complaints Against Various Television Licensees Concerning Their February 1, 2004, Broadcast of the Super Bowl XXXVIII Halftime Show*, File No. EB-04-IH-0011, FCC 06-19 (March 15, 2006).

[79] CBS Corp. v. Federal Communications Commission, No. 06-2209 (3d Cir.).

[80] *In the Matter of Complaints Regarding Various Television Broadcasts Between February 2, 2002 and March 8, 2005*, FCC 06-17 (March 15, 2006).

[81] *Id*. Also, three of the programs for which forfeitures were proposed were Spanish-language programs.

[82] The Commission found violations, but declined to impose forfeitures with respect to several programs that were aired prior to the Golden Globe Awards Order, at a time

when the Commission would not have taken enforcement actions against the isolated use of expletives. *Id.* at 100 - 137.

[83] For a complete list of recent actions related to broadcast indecency, see [http://www.fcc.gov/eb/broadcast/obscind.html].

[84] *In the Matter of Infinity Broadcasting, et al.*, EB-02-IH-0685 (October 2, 2003).

[85] *Id.* at 8.

[86] *Id.* at 9. The Commission noted that the contest portion of the broadcast in question lasted over an hour and was reproduced in a 203-page transcript.

[87] *See In the Matter of Viacom Inc., Infinity Radio Inc., et. al.*, FCC 04-268 (November 23, 2004) [http://hraunfoss.fcc.gov/edocs_public/attachmatch/FCC-04-268A1.pdf]. The decree also covers several other actions pending against Viacom-owned Infinity Radio stations and broadcast television stations, but does not cover the proceedings related to the Super Bowl halftime show discussed *supra*.

[88] *In the Matter of Clear Channel Broadcasting Licenses, Inc., et al.*, File No. EB-02-IH-0261 (January 27, 2004).

[89] *Id.* at 4.

[90] *Id.* at 5.

[91] *Id.* at 6.

[92] *Id.*

[93] *In the Matter of Clear Channel Broadcasting Licensees*, File No. EB-03-IH — 159 (April 8, 2004).

[94] *See* Statement of Commissioner Michael J. Copps, [http://hraunfoss.fcc.gov/edocs_public/attachmatch/DOC-245911A1.pdf], p. 2.

[95] *See In the Matter of Clear Channel Communications, Inc.*, FCC 04-128 (June 9, 2004) at [http://hraunfoss.fcc.gov/edocs_public/attachmatch/FCC-04-128A1.pdf].

[96] *Id.* at 7.

[97] Larry Neumeister, *Appeals court panel grills government lawyer in indecency case*, Associated Press State and Local Wire (December 20, 2006).

[98] Fox Television Stations, Inc. v. Federal Communications Commission, 489 F.3d 444, 447 (2d Cir. 2007).

[99] *Id.* at 458.

[100] *Id.*

[101] *Id.* at 459.

[102] *Id.*

[103] See text accompanying notes 67-70, *supra*.

[104] *Id.* at 462.

[105] This bill appears to be substantially similar to H.R. 3717, as reported by the House Committee on Energy and Commerce, on March 3, 2004.

[106] H.Amdt. 10. The FCC's policy statement on broadcast indecency was released on April 6, 2001, and can be found at [http://www.fcc.gov/eb/Orders/2001/fcc01090.html].

[107] See additional discussion on page 7, above.

[108] 438 U.S. 726 (1978). The FCC's action was to issue "a declaratory order granting the complaint," and "state that the order would be 'associated with the station's license file,'" which means that the FCC could consider it when it came time for the station's license renewal. *Id.* at 730.

[109] 521 U.S. 844, 868 (1997) (noting that "the history of the extensive regulation of the broadcast medium" and "the scarcity of available frequencies" are factors "not present in cyberspace," and striking down parts of the Communications Decency Act of 1996). The Court also cited *Pacifica* with approval in *United States v. Playboy Entertainment Group, Inc.*, 529 U.S. 803, 813-814 (2000), and in *Ashcroft v. Free Speech Coalition*, 535 U.S. 234, 245 (2002).

[110] *Pacifica, supra,* 438 U.S. at 750. A federal court of appeals subsequently held unconstitutional a federal statute that banned "indecent" broadcasts 24 hours a day, but, in a later case, the same court upheld the present statute, 47 U.S.C. § 303 note, which bans "indecent" broadcasts from 6 a.m. to 10 p.m. Action for Children's Television v. FCC, 932 F.2d 1504 (D.C. Cir. 1991), *cert. denied,* 503 U.S. 913 (1992); Action for Children's Television v. FCC, 58 F.3d 654 (D.C. Cir. 1995) (en banc), *cert. denied,* 516 U.S. 1043 (1996).

[111] *Id.* at 750, n.29.

[112] *Id.* at 746 n.22. These two sentences and the text accompanying the next footnote, although part of Justice Stevens' opinion, are in a part of the opinion (IV-B) joined by only two other justices. Every other quotation from *Pacifica* in this report was from a part of the opinion that a majority of the justices joined.

[113] *Id.* at 746.

[114] There also appears to be some tension between, on the one hand, Justice Stevens' distinction in *Pacifica* between a point of view and the way in which it is expressed, and, on the other hand, the Court's statement in *Cohen v. California* "that much linguistic expression serves a dual communicative function: it conveys not only ideas capable of relatively precise, detached explication, but otherwise inexpressible emotions as well. In fact, words are often chosen as much for their emotive as their cognitive force. We cannot sanction the view that the Constitution, while solicitous of the cognitive content of individual speech, has little or no regard for that emotive function which, practically speaking, may often be the more important element of the overall message sought to be communicated." 403 U.S. 15, 26 (1971) (upholding the First Amendment right, in the corridor of a courthouse, to wear a jacket bearing the words "F[***] the Draft"). Arguably, Carlin's use of "indecent" words not only served an emotive purpose, but served to indicate the precise words to whose censorship he was objecting. Yet *Pacifica* was decided after *Cohen,* which suggests that *Cohen* does not lessen the precedential value of *Pacifica.*

[115] Miller v. California, 413 U.S. 15, 27, 24 (1973). In addition, in striking down parts of the Communications Decency Act of 1996, the Court expressed concern that the statute may "extend to discussions about prison rape or safe sexual practices, artistic images that include nude subjects, and arguably the card catalogue of the Carnegie Library." *Reno v. ACLU, supra,* 521 U.S. at 878. And, in striking down a federal statute that prohibited child pornography that was produced without the use of an actual child, the Court expressed concern that the statute "prohibits speech despite its serious literary, artistic, political, or scientific value." Ashcroft v. Free Speech Coalition, 535 U.S. 234, 246 (2002). In neither of these cases, however, did the Court state that its holding turned on the statute's application to works of serious value.

[116] 390 U.S. 629 (1968).

[117] *Pacifica, supra,* 438 U.S. at 731-732.

[118] *Id.* at 732 (distinguishing "indecent" from "obscene" and "profane" in 18 U.S.C. § 1464).

[119] *Id.* at 740.

[120] *Id.* at 748-750.

[121] 95 U.S. 367, 369 (1969).

[122] *Id.* at 388.

[123] *Id.* at 394.

[124] *Id.* at 396.

[125] 512 U.S. 622, 639 (1994). In *Turner*, the Court held that the "must carry" rules, which "require cable television systems to devote a portion of their channels to the transmission of local broadcast television stations," *id.* at 626, were content-neutral and therefore not subject to strict scrutiny. The Court remanded and ultimately upheld the rules. Turner Broadcasting System, Inc., 520 U.S. 180 (1997).

[126] In the court of appeals decision upholding the current statute that bans "indecent" broadcasts from 6 a.m. to 10 p.m., a dissenting judge wrote of "the utterly irrational distinction that Congress has created between *broadcast* and *cable* operators. No one disputes that cable exhibits more and worse indecency than does broadcast. And cable television is certainly pervasive in our country." Action for Children's Television v. FCC, *supra*, 58 F.3d at 671 (emphasis in original) (Edwards, C.J., dissenting).

[127] 512 U.S. at 638 (citation omitted).

[128] Syracuse Peace Council v. FCC, 867 F.2d 654 (D.C. Cir. 1989), *cert. denied*, 493 U.S. 1019 (1990).

[129] *Id.* at 669. In *Arkansas AFL-CIO v. FCC*, 11 F.3d 1430 (8th Cir. 1993) (en banc), the court of appeals held that Congress had not codified the fairness doctrine and that the FCC's decision to eliminate it was a reasonable interpretation of the statutory requirement that licensees operate in the public interest.

[130] 518 U.S. 727, 748 (1996). The plurality added that cable television "is as 'accessible to children' as over-the-air broadcasting, if not more so," has also "established a uniquely pervasive presence in the lives of all Americans," and can also "'confron[t] the citizen' in 'the privacy of the home,' . . . with little or no prior warning." *Id.* at 744-745. Justice Souter concurred that "today's plurality opinion rightly observes that the characteristics of broadcast radio that rendered indecency particularly threatening in *Pacifica*, that is, its intrusion into the house and accessibility to children, are also present in the case of cable television. . . ." *Id.* at 776.

[131] 529 U.S. 803, 813 (2000) (striking down a federal statute that required distributors to fully scramble or fully block signal bleed to non-subscribers to cable channels; "signal bleed" refers to the audio or visual portions of cable television programs that nonsubscribers to a cable channel may be able to hear or see despite the fact that the programs have been scrambled to prevent the non-subscribers from hearing or seeing them).

[132] An earlier district court case held that *Pacifica* does not apply to cable television because of several differences between cable and broadcasting. For one, "[i]n the cable medium, the physical scarcity that justifies content regulation in broadcasting is not present." For another, as a subscriber medium, "cable TV is not an intruder but an invitee whose invitation can be carefully circumscribed." Community Television v. Wilkinson, 611 F. Supp. 1099 (D. Utah 1985), *aff'd*, 800 F.2d 989 (10th Cir. 1986),

aff'd, 480 U.S. 926 (1987) (striking down Utah Cable Television Programming Decency Act). The court of appeals did not discuss the constitutional issue beyond stating that it agreed with the district court's reasons for its holding. 800 F.2d at 991. A summary affirmance by the Supreme Court, as in this case, is "an affirmance of the judgment only," and does not indicate approval of the reasoning of the court below. Mandel v. Bradley, 432 U.S. 173, 176 (1977).

[133] See CRS Report RL33170, *Constitutionality of Applying the FCC's Indecency Restriction to Cable Television*, by Henry Cohen, which concludes that "it appears likely that a court would find that to apply the FCC's indecency restriction to cable television would be unconstitutional."

[134] Sable Communications of California v. Federal Communications Commission, 492 U.S. 115 (1989); Action for Children's Television v. FCC, *supra*, 932 F.2d at 1509.

[135] *Id.* at 126.

[136] Cohen v. California, *supra*, 403 U.S. at 19, 21. Under *Pacifica*, broadcast media do thrust words upon unwilling or unsuspecting viewers, but, if a court were to apply strict scrutiny to a ban on the broadcast of "indecent" language regardless of context, then it would not be following *Pacifica*.

[137] *Turner Broadcasting, supra*, 512 U.S. at 664 (incidental restriction on speech). *See also*, Edenfield v. Fane, 507 U.S. 761, 770-771 (1993) (restriction on commercial speech); Nixon v. Shrink Missouri Government PAC, 528 U.S. 377, 392 (2000) (restriction on campaign contributions). In all three of these cases, the government had restricted less-than-fully protected speech, so the Court did not apply strict scrutiny. Because offensive words are apparently entitled to full First Amendment protection (except under *Pacifica* and in the instances cited in *Cohen v. California*, quoted in the text above), it seems all the more likely that the Court, if it applied strict scrutiny instead of *Pacifica* to a challenge to a ban on the broadcast of "indecent" words regardless of context, would require the government to demonstrate that harms it recites are real and that the ban would alleviate these harms in a direct and material way.

[138] Interactive Digital Software Association v. St. Louis County, Missouri, 329 F.3d 954, 959 (8th Cir. 2003). The Supreme Court has "recognized that there is a compelling interest in protecting the physical and psychological well-being of minors. This interest extends to shielding minors from the influence of literature that is not obscene by adult standards." *Sable, supra*, 492 U.S. at 126. The Court has also upheld a state law banning the distribution to minors of "so-called 'girlie' magazines" even as it acknowledged that "[i]t is very doubtful that this finding [that such magazines are "a basic factor in impairing the ethical and moral development of our youth"] expresses an accepted scientific fact." Ginsberg v. New York, *supra*, 390 U.S. at 631, 641. "To sustain state power to exclude [such material from minors]," the Court wrote, "requires only that we be able to say that it was not irrational for the legislature to find that exposure to material condemned by the statute is harmful to minors." *Id.* at 641. *Ginsberg* thus "invokes the much less exacting 'rational basis' standard of review," rather than strict scrutiny. *Interactive Digital Software Association, supra*, 329 F.3d at 959. A federal district court wrote: We are troubled by the absence of evidence of harm presented both before Congress and before us that the viewing of signal bleed of sexually explicit programming causes harm to children and that the avoidance of this harm can be recognized as a compelling State interest. We recognize that the Supreme

Court's jurisprudence does not require empirical evidence. Only some minimal amount of evidence is required when sexually explicit programming and children are involved. Playboy Entertainment Group, Inc. v. United States, 30 F. Supp.2d 702, 716 (D. Del. 1998), *aff'd*, 529 U.S. 803 (2000). The district court therefore found that the statute served a compelling governmental interest, though it held it unconstitutional because it found that the statute did not constitute the least restrictive means to advance the interest. The Supreme Court affirmed on the same ground, apparently assuming the existence of a compelling governmental interest, but finding a less restrictive means that could have been used. In another case, a federal court of appeals, upholding the current statute that bans "indecent" broadcasts from 6 a.m. to 10 p.m., noted "that the Supreme Court has recognized that the Government's interest in protecting children extends beyond shielding them from physical and psychological harm. The statute that the Court found constitutional in *Ginsberg* sought to protect children from exposure to materials that would 'impair [their] *ethical and moral* development. . . . Congress does not need the testimony of psychiatrists and social scientists in order to take note of the coarsening of impressionable minds that can result from a persistent exposure to sexually explicit material. . . .'" Action for Children's Television v. FCC, *supra*, 58 F.3d at 662 (brackets and italics supplied by the court). A dissenting judge in the case noted that, "[t]here is not one iota of evidence in the record . . . to support the claim that exposure to indecency is harmful — indeed, the nature of the alleged 'harm' is never explained." *Id.* at 671 (D.C. Cir. 1995) (Edwards, C.J., dissenting).

[139] The full Commission's decision in the Bono case stated that "any use of that word or a variation, in any context, inherently has a sexual connotation." But this does not necessarily mean that it is sexually oriented enough to cause the courts to assume without evidence that it is harmful to minors.

[140] *Ginsberg, supra*, 390 at 640, 639. *See also*, Action for Children's Television v. FCC, *supra*, 58 F.3d at 661.

[141] Action for Children's Television v. FCC, *supra*, 58 F.3d at 670 (emphasis in original) (Edwards, C.J., dissenting).

[142] *Pacifica, supra*, 438 U.S. at 732.

[143] *Fox Television Stations, supra* note 98, 489 F.3d at 447.

[144] *Id.* at 463-465.

In: Telecommunications and Media Issues
Editors: A. N. Moller, C. E. Pletson, pp. 33-50

ISBN: 978-1-60456-294-1
© 2008 Nova Science Publishers, Inc.

.

Chapter 2

"SPAM": AN OVERVIEW OF ISSUES CONCERNING COMMERCIAL ELECTRONIC MAIL[*]

Patricia Moloney Figliola

ABSTRACT

Spam, also called unsolicited commercial email (UCE) or "junk email," aggravates many computer users. Not only can spam be a nuisance, but its cost may be passed on to consumers through higher charges from Internet service providers who must upgrade their systems to handle the traffic. Also, some spam involves fraud, or includes adult-oriented material that offends recipients or that parents want to protect their children from seeing. Proponents of UCE insist it is a legitimate marketing technique that is protected by the First Amendment, and that some consumers want to receive such solicitations.

On December 16, 2003, President Bush signed into law the Controlling the Assault of Non-Solicited Pornography and Marketing (CAN-SPAM) Act, P.L. 108-187. It went into effect on January 1, 2004. The CAN-SPAM Act does not ban UCE. Rather, it allows marketers to send commercial email as long as it conforms with the law, such as including a legitimate opportunity for consumers to "opt-out" of receiving future commercial emails from that sender. It preempts state laws that specifically address spam, but not state laws that are not specific to email, such as trespass, contract, or tort law, or other state laws to the extent they relate to fraud or computer crime. It does not require a centralized "Do Not Email" registry to be created by the Federal Trade Commission (FTC), similar to the National Do Not Call registry for telemarketing. The law requires only that the FTC develop a plan and timetable for establishing such a registry, and to inform Congress of any concerns it has with regard to establishing it. The FTC submitted a report to Congress on June 15, 2004, concluding that a Do Not Email registry could actually increase spam.

Proponents of CAN-SPAM have argued that consumers are most irritated by *fraudulent* email, and that the law should reduce the volume of such email because of the civil and criminal penalties included therein. Opponents counter that consumers object to *unsolicited* commercial email, and since the law legitimizes commercial email (as long as it conforms with the law's provisions), consumers actually may receive more, not fewer, UCE messages. Thus, whether or not "spam" is reduced depends in part on whether it is

[*] Excerpted from CRS Report RL31953, dated October 10, 2007.

defined as only fraudulent commercial email, or all unsolicited commercial email. Many observers caution that consumers should not expect any law to solve the spam problem — that consumer education and technological advancements also are needed.

Note: This report was originally written by Marcia S. Smith; the author acknowledges her contribution to CRS coverage of this issue area.

INTRODUCTION

One aspect of increased use of the Internet for electronic mail (e-mail) has been the advent of unsolicited advertising, also called "unsolicited commercial e-mail" (UCE), "unsolicited bulk e-mail," "junk e-mail, "or "spam."[1] Complaints often focus on the fact that some spam contains, or has links to, pornography; that much of it is fraudulent; and the volume of spam is steadily increasing.[2] However, recent research shows that Internet users' concerns about spam are actually decreasing, even while the volume of spam continues to increase. For example, in a survey conducted by the Pew Internet and American Life Project during February and March 2007, respondents stated that they were "less bothered by [spam]" now than they reported being in the previous survey, conducted in June 2003. Specifically, in the 2003 survey, 25% of respondents stated that spam was a "big problem"; in the 2007 survey, that figure had dropped to 18%. Even more striking is that the percentage of participants who responded that spam was "not a problem at all" rose from 16% to 28% between 2003 and 2007. The percentage of respondents stating that spam is "an annoyance, but not a big problem" has stayed roughly the same at 57% and 51% in 2003 and 2007, respectively.[3]

One reason for this change in attitude towards spam is attributed to Internet users' growing savvy with identifying spam on their own as well as their increased use of spam filters (whether provided by their Internet service provider (ISP) or purchased on their own). Currently, 71% of Internet users use filters, up from 65% in 2005.[4]

Opponents of junk e-mail argue that not only is it annoying and an invasion of privacy,[5] but that its cost is borne by recipients and ISPs, not the marketers. Consumers reportedly are charged higher fees by ISPs that must invest resources to upgrade equipment to manage the high volume of e-mail, deal with customer complaints, and mount legal challenges to junk e-mailers. Businesses may incur costs due to lost productivity, or investing in upgraded equipment or anti-spam software. The Ferris Research Group,[6] which offers consulting services on managing spam, estimated in 2003 that spam cost U.S. organizations over $10 billion.

Proponents of UCE argue that it is a valid method of advertising, and is protected by the First Amendment. The Direct Marketing Association (DMA) released figures in May 2003 showing that commercial e-mail generates more than $7.1 billion in annual sales and $1.5 billion in potential savings to American consumers.[7] In a joint open letter to Congress published in *Roll Call* on November 13, 2003, three marketing groups — DMA, the American Association of Advertising Agencies, and the Association of National Advertisers — asserted that "12% of the $138 billion Internet commerce marketplace is driven by legitimate commercial e-mail. This translates into a minimum of $17.5 billion spent in response to commercial e-mails in 2003 for bedrock goods and services such as travel, hotels, entertainment, books, and clothing." A March 2004 study by the Pew Internet and American

Life Project found that 5% of e-mail users said they had ordered a product or service based on an unsolicited e-mail, which "translates into more than six million people."[8]

DMA argued for several years that instead of banning UCE, individuals should be given the opportunity to "opt-out" by notifying the sender that they want to be removed from the mailing list. (The concepts of opt-out and opt-in are discussed below.) Hoping to demonstrate that self regulation could work, in January 2000, the DMA launched the E-mail Preference Service where consumers who wish to opt-out can register themselves at a DMA website.[9] DMA members sending UCE must check their lists of recipients and delete those who have opted out. Critics argued that most spam does not come from DMA members, so the plan was insufficient, and on October 20, 2002, the DMA agreed. Concerned that the volume of unwanted and fraudulent spam is undermining the use of e-mail as a marketing tool, the DMA announced that it would pursue legislation to battle the rising volume of spam.

DEFINING SPAM

One challenge in debating the issue of spam is defining it.[10] To some, it is any commercial e-mail to which the recipient did not "opt-in" by giving prior *affirmative consent* to receiving it. To others, it is commercial e-mail to which *affirmative* or *implied consent* was not given, where implied consent can be defined in various ways (such as whether there is a pre-existing business relationship). Still others view spam as "unwanted" commercial e-mail. Whether or not a particular e-mail is unwanted, of course, varies per recipient. Since senders of UCE do find buyers for some of their products, it can be argued that at least some UCE is reaching interested consumers, and therefore is wanted, and thus is not spam. Consequently, some argue that marketers should be able to send commercial e-mail messages as long as they allow each recipient an opportunity to indicate that future such e-mails are not desired (called "opt-out"). Another group considers spam to be only fraudulent commercial e-mail, and believe that commercial e-mail messages from "legitimate" senders should be permitted. The DMA, for example, considers spam to be only fraudulent UCE.

The differences in defining spam add to the complexity of devising legislative or regulatory remedies for it. Some of the bills introduced in the 108th Congress took the approach of defining commercial e-mail, and permitting such e-mail to be sent to recipients as long as it conformed with certain requirements. Other bills defined *unsolicited* commercial e-mail and prohibited it from being sent unless it met certain requirements. The final law, the CAN-SPAM Act (see below), took the former approach, defining and allowing marketers to send such e-mail as long as they abide by the terms of the law, such as ensuring that the e-mail does not have fraudulent header information or deceptive subject headings, and includes an opt-out opportunity and other features that proponents argue will allow recipients to take control of their in-boxes. Proponents of the law argue that consumers will benefit because they should see a reduction in fraudulent e-mails. Opponents of the law counter that it legitimizes sending commercial e-mail, and to the extent that consumers do not want to receive such e-mails, the amount of unwanted e-mail actually may increase. If the legislation reduces the amount of fraudulent e-mail, but not the amount of unwanted e-mail, the extent to which it reduces "spam" would depend on what definition of that word is used.

On December 16, 2004, the FTC issued its final rule defining the term "commercial electronic mail message," but explicitly declined to define "spam."

.

AVOIDING AND REPORTING SPAM

Tips on avoiding spam are available on the FTC website[11] and from Consumers Union.[12] Consumers may file a complaint about spam with the FTC by visiting the FTC website and choosing "File a Complaint" at the bottom of the page.[13] The offending spam also may be forwarded to the FTC, at spam@uce.gov, to assist the FTC in monitoring spam trends and developments. The September 2004 issue of *Consumer Reports* has a cover story about spam, including ratings of commercially available spam filters consumers can load onto their computers. Also, individual ISPs use spam filters (though the filters may not catch all spam) and have mechanisms available for subscribers to report spam.

Foreign Spam

Controlling spam is complicated by the fact that some of it originates outside the United States and thus is not subject to U.S. laws or regulations. Spam is a global problem, and a 2001 study by the European Commission concluded that Internet subscribers globally pay 10 billion Euros a year in connection costs to download spam.[14] Some European officials complain that the United States is the source of most spam, and the U.S. decision to adopt an opt-out approach in the CAN-SPAM Act (discussed below) was not helpful.[15] In April 2005, a British anti-spam and anti-virus software developing company, Sophos, listed the United States as the largest spam producing country, exporting 35.7% of spam (down from 42.1% in December 2004); South Korea was second, at 25% (up from 13.4% in December 2004).[16] Tracing the origin of any particular piece of spam can be difficult because some spammers route their messages through other computers (discussed below) that may be located anywhere on the globe.

THE FEDERAL CAN-SPAM ACT: SUMMARY OF MAJOR PROVISIONS

The 108th Congress passed the CAN-SPAM Act, S. 877, which merged provisions from several House and Senate bills.[17] Signed into law by President Bush on December 16, 2003 (P.L. 108-187), it went into effect on January 1, 2004.[18] P.L. 108-187 includes the following major provisions.

Commercial e-mail may be sent to recipients as long as the message conforms with the following requirements:

- transmission information in the header is not false or misleading;
- subject headings are not deceptive;

- a functioning return e-mail address or comparable mechanism is included to enable recipients to indicate they do not wish to receive future commercial e-mail messages from that sender at the e-mail address where the message was received;
- the e-mail is not sent to a recipient by the sender, or anyone acting on behalf of the sender, more than 10 days after the recipient has opted-out, unless the recipient later gives affirmative consent to receive the e-mail (i.e., opts back in); and
- the e-mail must be clearly and conspicuously identified as an advertisement or solicitation (although the legislation does not state how or where that identification must be made).

- Commercial e-mail is defined as e-mail, the primary purpose of which is the commercial advertisement or promotion of a commercial product or service (including content on an Internet website operated for a commercial purpose). It does not include transactional or relationship messages (see next bullet). The act directs the FTC to issue regulations within 12 months of enactment to define the criteria to facilitate determination of an e-mail's primary purpose. The FTC did so on December 16, 2004.

- Some requirements (including the prohibition on deceptive subject headings, and the opt-out requirement) do not apply if the message is a "transactional or relationship message," which include various types of notifications, such as periodic notifications of account balance or other information regarding a subscription, membership, account, loan or comparable ongoing commercial relationship involving the ongoing purchase or use by the recipient of products or services offered by the sender; providing information directly related to an employment relationship or related benefit plan in which the recipient is currently involved, participating, or enrolled; or delivering goods or services, including product updates or upgrades, that the recipient is entitled to receive under the terms of a transaction that the recipient has previously agreed to enter into with the sender. The act allows, but does not require, the FTC to modify that definition.

- Sexually-oriented commercial e-mail must include, in the subject heading, a "warning label" to be prescribed by the FTC (in consultation with the Attorney General), indicating its nature. The warning label does not have to be in the subject line, however, if the message that is initially viewable by the recipient does not contain the sexually oriented material, but only a link to it. In that case, the warning label, and the identifier, opt-out, and physical address required under section 5 (a)(5) of the act; must be contained in the initially viewable e-mail message as well. Sexually oriented material is defined as any material that depicts sexually explicit conduct, unless the depiction constitutes a small and insignificant part of the whole, the remainder of which is not primarily devoted to sexual matters. These provisions do not apply, however, if the recipient has given prior affirmative consent to receiving such e-mails.

- Businesses may not knowingly promote themselves with e-mail that has false or misleading transmission information.

- State laws specifically related to spam are preempted, but not other state laws that are not specific to electronic mail, such as trespass, contract, or tort law, or other state laws to the extent they relate to fraud or computer crime.
- Violators may be sued by FTC, state attorneys general, and ISPs (but not by individuals).
- Violators of many of the provisions of the act are subject to statutory damages of up to $250 per e-mail, to a maximum of up to $2 million, which may be tripled by the court (to $6 million) for "aggravated violations."
- Violators may be fined, or sentenced to up to 3 or five years in prison (depending on the offense), or both, for accessing someone else's computer without authorization and using it to send multiple commercial e-mail messages; sending multiple commercial e-mail messages with the intent to deceive or mislead recipients or ISPs as to the origin of such messages; materially falsifying header information in multiple commercial e-mail messages; registering for five or more e-mail accounts or online user accounts, or two or more domain names, using information that materially falsifies the identity of the actual registrant, and sending multiple commercial e-mail messages from any combination of such accounts or domain names; or falsely representing oneself to be the registrant or legitimate successor in interest to the registrant of five of more Internet Protocol addresses, and sending multiple commercial e-mail messages from such addresses. "Multiple" means more than 100 e-mail messages during a 24-hour period, more than 1,000 during a 30-day period, or more than 10,000 during a one-year period. Sentencing enhancements are provided for certain acts.
- The Federal Communications Commission, in consultation with the FTC, must prescribe rules to protect users of wireless devices from unwanted commercial messages. (The rules were issued in August 2004. See CRS Report RL31636, *Wireless Privacy and Spam: Issues for Congress*, by Marcia S. Smith, for more on this topic.)

Conversely, the act does not —

- Create a "Do Not Email registry" where consumers can place their e-mail addresses in a centralized database to indicate they do not want commercial e-mail. The law required only that the FTC develop a plan and timetable for establishing such a registry and to inform Congress of any concerns it has with regard to establishing it. (The FTC released that report in June 2004; see next section.)
- Require that consumers "opt-in" before receiving commercial e-mail.
- Require commercial e-mail to include an identifier such as "ADV" in the subject line to indicate it is an advertisement. The law does require the FTC to report to Congress within 18 months of enactment on a plan for requiring commercial e-mail to be identifiable from its subject line through use of "ADV" or a comparable identifier, or compliance with Internet Engineering Task Force standards, or an explanation of any concerns FTC has about such a plan.
- Include a "bounty hunter" provision to financially reward persons who identify a violator and supply information leading to the collection of a civil penalty, although

the FTC must submit a report to Congress within nine months of enactment setting forth a system for doing so. (The study was released in September 2004.)

OPT-IN, OPT-OUT, AND A "DO NOT EMAIL" REGISTRY

Much of the debate on how to stop spam focuses on whether consumers should be given the opportunity to "opt-in" (where prior consent is required) or "opt-out" (where consent is assumed unless the consumer notifies the sender that such e-mails are not desired) of receiving UCE or all commercial e-mail. The CAN-SPAM Act is an "opt out" law, requiring senders of all commercial e-mail to provide a legitimate[19] opt-out opportunity to recipients.

During debate on the CAN-SPAM Act, several anti-spam groups argued that the legislation should go further, and prohibit commercial e-mail from being sent to recipients unless they opt-in, similar to a policy adopted by the European Union (see below). Eight U.S. groups, including Junkbusters, the Coalition Against Unsolicited Commercial Email (CAUCE), and the Consumer Federation of America, wrote a letter to several Members of Congress expressing their view that the opt-out approach (as in P.L. 108-187) would "undercut those businesses who respect consumer preferences and give legal protection to those who do not."[20] Some of the state laws (see below) adopted the opt-in approach, including California's anti-spam law.

The European Union adopted an opt-in requirement for e-mail, which became effective October 31, 2003.[21] Under the EU policy, prior affirmative consent of the recipient must be obtained before sending commercial e-mail unless there is an existing customer relationship. In that case, the sender must provide an opt-out opportunity. The EU directive sets the broad policy, but each member nation must pass its own law as to how to implement it.[22]

As noted, Congress chose opt-out instead of opt-in, however. One method of implementing opt-out is to create a "Do Not Email" registry where consumers could place their names on a centralized list to opt-out of all commercial e-mail instead of being required to respond to individual e-mails. The concept is similar to the National Do Not Call registry where consumers can indicate they do not want to receive telemarketing calls. During consideration of the CAN-SPAM Act, then-FTC Chairman Timothy Muris and other FTC officials repeatedly expressed skepticism about the advisability of a Do Not Email registry despite widespread public support for it.[23] One worry is that the database containing the e-mail addresses of all those who do not want spam would be vulnerable to hacking, or spammers otherwise might be able to use it to obtain the e-mail addresses of individuals who explicitly do not want to receive spam. In an August 19, 2003, speech to the Aspen Institute, Mr. Muris commented that the concept of a Do Not Email registry was interesting, "but it is unclear how we can make it work" because it would not be enforceable.[24] "If it were established, my advice to consumers would be: Don't waste the time and effort to sign up."

Following initial Senate passage of S. 877, an unnamed FTC official was quoted by the *Washington Post* as saying that the FTC's position on the registry is unchanged, and "Congress would have to change the law" to require the FTC to create it.[25] After the House passed S. 877, Mr. Muris released a statement complimenting Congress on taking a positive step in the fight against spam, but cautioned again that legislation alone will not solve the problem.[26]

CAN-SPAM Act Provision

The CAN-SPAM Act did not require the FTC to create a Do Not Email registry.[27] Instead, it required the FTC to submit a plan and timetable for establishing a registry, authorized the FTC to create it, and instructed the FTC to explain to Congress any concerns about establishing it.

FTC Implementation

The FTC issued its report to Congress on June 15, 2004.[28] The report concluded that without a technical system to authenticate the origin of e-mail messages, a Do Not Email registry would not reduce the amount of spam, and, in fact, might increase it.

The FTC report stated that "spammers would most likely use a Registry as a mechanism for verifying the validity of e-mail addresses and, without authentication, the Commission would be largely powerless to identify those responsible for misusing the Registry. Moreover, a Registry-type solution to spam would raise serious security, privacy, and enforcement difficulties." (p. I) The report added that protecting children from "the Internet's most dangerous users, including pedophiles," would be difficult if the Registry identified accounts used by children in order to assist legitimate marketers from sending inappropriate messages to them. (p. I) The FTC described several registry models that had been suggested, and computer security techniques that some claimed would eliminate or alleviate security and privacy risks. The FTC stated that it carefully examined those techniques — a centralized scrubbing of marketers' distribution lists, converting addresses to one-way hashes (a cryptographic approach), and seeding the Registry with "canary" e-mail addresses — to determine if they could effectively control the risks "and has concluded that none of them would be effective." (p. 16)

The FTC concluded that a necessary prerequisite for a Do Not Email registry is an authentication system that prevents the origin of e-mail messages from being falsified, and proposed a program to encourage the adoption by industry of an authentication standard. If a single standard does not emerge from the private sector after a sufficient period of time, the FTC report said the Commission would initiate a process to determine if a federally mandated standard is required. If the government mandates a standard, the FTC would then consider studying whether an authentication system, coupled with enforcement or other mechanisms, had substantially reduced the amount of spam. If not, the Commission would then reconsider whether or not a Do Not Email registry is needed.

On August 1, 2005, the FTC issued a press release summarizing the results of testing it had conducted to determine if online retailers were honoring opt-out requests. The FTC found that 89% of the merchants it tested did, in fact, stop sending e-mails when requested to do so.[29]

LABELS

Another approach to restraining spam is requiring that senders of commercial e-mail use a label, such as "ADV," in the subject line of the message, so the recipient will know before opening an e-mail message that it is an advertisement. That would also make it easier for spam filtering software to identify commercial e-mail and eliminate it. Some propose that adult-oriented spam have a special label, such as ADV-ADLT, to highlight that the e-mail may contain material or links that are inappropriate for children, such as pornography.

CAN-SPAM Act Provision

The CAN-SPAM Act: (1) requires clear and conspicuous identification that a commercial e-mail is an advertisement, but is not specific about how or where that identification must be made; (2) requires the FTC to prescribe warning labels for sexually-oriented e-mails within 120 days of enactment; and (3) requires the FTC to submit a report within 18 months of enactment setting forth a plan for requiring commercial e-mail to be identifiable from its subject line using ADV or a comparable identifier, or by means of compliance with Internet Engineering Task Force standards. However, the clear and conspicuous identification that a commercial e-mail is an advertisement, and the warning label for sexually-oriented material, are not required if the recipient has given prior affirmative consent to receipt of such messages.

FTC Implementation

On May 19, 2004, an FTC rule regarding labeling of sexually oriented commercial e-mail went into effect. The rule was adopted by the FTC (5-0) on April 13, 2004. A press release and the text of the ruling are available on the FTC's website.[30] The rule requires that the mark "SEXUALLY-EXPLICIT" be included both in the subject line of any commercial e-mail containing sexually oriented material, and in the body of the message in what the FTC called the "electronic equivalent of a 'brown paper wrapper.'" The FTC explained that the "brown paper wrapper" is what a recipient initially sees when opening the e-mail, and it may not contain any other information or images except what the FTC prescribes. The rule also clarifies that the FTC interprets the CAN-SPAM Act provisions to include both visual images and written descriptions of sexually explicit conduct.

On July 20, 2005, the FTC announced that it had charged seven companies with violating federal laws requiring these labels. Four of the companies settled with the FTC, which imposed a total of $1.159 million in civil penalties. U.S. District Court suits were filed against the other three companies.[31]

The act also required the FTC to submit a report to Congress on a plan for making commercial e-mail identifiable from its subject line, or to explain what concerns would lead the FTC to recommend against such a plan. That report was submitted in June 2005. It concluded that requiring UCE senders to use a prefix such as ADV probably would not result in less spam.

Experience with subject line labeling requirements in the states and in other countries does not support the notion that such requirements are an effective means of reducing spam.... Indeed, spam filters widely available at little or no cost ... more effectively empower consumers to set individualized email preferences to reduce unwanted UCE from both spammers and legitimate marketers. Mandatory subject line labeling, by comparison, would be an imprecise tool ... that, at best, might make it easier to segregate *labeled* UCE from *unlabeled* UCE. ... [I]t is extremely unlikely that outlaw spammers would comply with a requirement to label the email messages they send. By contrast, legitimate marketers likely *would* comply.... As a result ... labeled UCE messages sent by law-abiding senders would be filtered out. Meanwhile, unlabeled UCE messages sent by outlaw spammers would still reach consumers' in-boxes.[32] (Italics in original.)

OTHER IMPLEMENTATION ACTIONS

The act required the FTC or the Federal Communications Commission (FCC) to take a number of other actions with regard to implementing the CAN-SPAM Act. The FTC routinely issues Notices of Proposed Rulemaking or the results thereof regarding this act, which are too numerous to include in this report. Selected issues are addressed below. See the FTC's spam website [http://www.ftc.gov/spam] for more information.

Wireless Spam

The act required the FCC to issue regulations concerning spam on wireless devices such as cell phones. The FCC issued those regulations in August 2004.[33]

"Bounty Hunter" Provision

The act required the FTC to conduct a study on whether rewarding persons who identify a spammer and supply information leading to the collection of a civil penalty could be an effective technique for controlling spam (the "bounty hunter" provision). The study was released on September 15, 2004.[34] The FTC concluded that the benefits of such a system are unclear because, for example, without large rewards (in the $100,000 to $250,000 range) and a certain level of assurance that they would receive the reward, whistleblowers might not be willing to assume the risks of providing such information. The FTC offered five recommendations if Congress wants to pursue such an approach:

- tie eligibility for a reward to imposition of a final court order, instead of to collecting a civil penalty;
- fund the rewards through congressional appropriations, instead of through collected civil penalties;
- restrict reward eligibility to insiders with high-value information;
- exempt FTC decisions on eligibility for rewards from judicial or administrative review; and

- establish reward amounts high enough to attract insiders with high-value information.

Definition of "Primary Purpose"

The act required the FTC to issue regulations, within one year of enactment, defining the relevant criteria to facilitate determination of an e-mail's "primary purpose." The FTC issued its final rule on December 16, 2004, exactly one year after the law was enacted. According to the FTC's press release, [35] the final rule clarifies that the Commission does not intend to regulate non-commercial speech. It differentiates between commercial content and transactional or relationship" content in defining the primary purpose of an e-mail message.

- If an e-mail contains only a commercial advertisement or promotion of a commercial product or service, its primary purpose is deemed to be commercial.
- If an e-mail contains both commercial content and transactional or relationship content, the primary purpose is deemed to be commercial if the recipient would likely conclude that it was commercial through reasonable interpretation of the subject line, or if the transactional and relationship content does not appear in whole or in substantial part at the beginning of the body of the message.
- If an e-mail contains both commercial content, and content that is neither commercial content nor transactional or relationship content, the primary purpose is deemed to be commercial if the recipient would likely conclude that it was commercial through reasonable interpretation of the subject line, or if the recipient would likely conclude the primary purpose was commercial through reasonable interpretation of the body of the message.
- If an e-mail contains only transactional or relationship content, it is not deemed to be a commercial e-mail message.

"Commercial" content is defined in the final rule as "the commercial advertisement or promotion of a commercial product or service," which includes "content on an Internet website operated for a commercial purpose." That is the same as the definition in the CAN-SPAM Act.[36]

The FTC specifically declined to define the term "spam" because the act sets forth a regulatory scheme built around the terms "commercial electronic mail message" and "transactional or relationship message."[37]

RELATED LEGISLATION

On December 22, 2006, President Bush signed the Undertaking Spam, Spyware, And Fraud Enforcement With Enforcers beyond Borders Act of 2005 (U.S. SAFE WEB Act, (P.L. 109-455). The law allows the FTC and parallel foreign law enforcement agencies to share information while investigating allegations of "unfair and deceptive practices" that involve foreign commerce, but raised some privacy concerns because the FTC would not be required to make public any of the information it obtained through foreign sources.

LEGAL ACTIONS BASED ON THE CAN-SPAM ACT

Many lawsuits have been brought against spammers. The following discussion is illustrative, not comprehensive.

On October 10, 2007, the FTC announced that it had filed a civil lawsuit against an international enterprise, with defendants in the United States, Canada, and Australia, that used spam to drive traffic to websites selling pills that the FTC alleges do not work.[38] The FTC's spam database received over 175,000 spam messages sent on behalf of the operation. The action, announced at an international meeting of government authorities and private industry about spam, spyware, and other online threats, is the first brought by the agency using the U.S. SAFE WEB Act to share information with foreign partners. In addition, the FTC alleges that the operation violated the CAN-SPAM Act by initiating commercial e-mails that contained false "from" addresses and deceptive subject lines, and failed to provide an opt-out link or physical postal address.

On April 29, 2004, the FTC announced that it had filed a civil lawsuit against a Detroit-based spam operation, Phoenix Avatar, and the Department of Justice (DOJ) announced that it had arrested two (and were seeking two more) Detroit-area men associated with the company who are charged with sending hundreds of thousands of spam messages using false and fraudulent headers.[39] The FTC charged Phoenix Avatar with making deceptive claims about a diet patch sold via the spam in violation of the FTC Act, and with violations of the CAN-SPAM Act because the spam did not contain a valid opt-out opportunity and the "reply to" and "from" addresses were fraudulent. The DOJ filed criminal charges against the men under the CAN-SPAM Act for sending multiple commercial e-mails with materially false or fraudulent return addresses. According to the FTC, from January 1, 2004 until the lawsuit was filed, about 490,000 of the spam messages forwarded by consumers to the FTC were linked to Avatar Phoenix.

The FTC simultaneously announced that it had filed a legal action against an Australian spam enterprise operating out of Australia and New Zealand called Global Web Promotions. The FTC stated that it was assisted by the Australian Competition and Consumer Commission and the New Zealand Commerce Committee in bringing the case. According to the FTC, since January 1, 2004, among the spam forwarded by consumers to the FTC, about 399,000 are linked to Global Web Promotions. The FTC charges that a diet patch, and human growth hormone products, sold by Global Web Promotions are deceptive and in violation of the FTC Act. The products are shipped from within the United States. The FTC further charges that the spam violates the CAN-SPAM Act because of fraudulent headers.

The FTC also filed a complaint against six companies and five individuals who, the FTC alleges, acting as a single business enterprise, sent e-mails containing sexually-explicit content without the required warning label and violated other provisions of the Adult Labeling Rule, the CAN-SPAM Act, and the FTC Act.[40] A federal district court issued a Temporary Restraining Order against the defendants.

Separately, four of the largest ISPs — AOL, Earthlink, Microsoft, and Yahoo! — working together as part of the Anti-Spam Alliance, filed civil suits under the CAN-SPAM Act against hundreds of alleged spammers in March 2004.[41] The suits were filed in federal courts in California, Georgia, Virginia and Washington. A number of other suits since have been filed.

The Massachusetts Attorney General filed the first state CAN-SPAM case against a Florida business called DC Enterprises, and its proprietor William T. Carson in July 2004, which also was filed under the Massachusetts Consumer Protection Act.[42] That case was settled by DC Enterprises and Mr. Carson, who agreed to pay $25,000, halt further violations of the CAN-SPAM Act, and comply with state regulations regarding mortgage brokers.[43]

It should be noted, however, that some ISPs are having difficulty recovering monetary judgments from spam cases (though not necessarily cases brought under the CAN-SPAM Act). Microsoft, for example, reportedly has won $620 million in judgments, but has collected only $500,000.[44]

FTC ACTIVITY

The FTC enforces the CAN-SPAM Act and conducts other consumer-education initiatives related to combating spam.

2005 Assessment of the CAN-SPAM Act

Under the law, the FTC was required to provide Congress with an assessment of the act's effectiveness, and recommend any necessary changes. The FTC submitted its report in December 2005.[45] The FTC concluded that the act has been effective in terms of adoption of commercial e-mail "best practices" that are followed by "legitimate" online marketers, and in terms of providing law enforcement agencies and ISPs with an additional tool to use against spammers. Additionally, the FTC concluded that the volume of spam has begun to stabilize, and the amount reaching individuals' inboxes has decreased because of improved anti-spam technologies.[46] However, it also found that the international dimension of spam has not changed significantly, and that there has been a shift toward the inclusion of "increasingly malicious" content in spam messages, such as "malware," which is intended to harm the recipient. Other negative changes noted by the FTC are that spammers are using increasingly complex multi-layered business arrangements to frustrate law enforcement, and are hiding their identities by providing false information to domain registrars (the "Whois" database).

The FTC did not recommend any changes to the CAN-SPAM Act, but encouraged Congress to pass the US SAFE WEB Act (S. 1608, see next paragraph), noted that continued consumer education efforts are needed, and called for improved anti-spam technologies, particularly domain-level authentication (discussed later in this report).

July 2007 Spam Summit

In July 2007, the FTC hosted "Spam Summit: The Next Generation of Threats and Solutions."[47] This event was a follow-on effort of the FTC's 2003 Spam Forum. Issues included defining the problem; new methods for sending spam; the "covert economy" (e.g., to what extent does stolen information, such as government-issued identity numbers, credit cards, bank cards and personal identification numbers, user accounts, and e-mail addresses, play a role in spam?); deterring malicious spammers and cybercriminals; emerging threats (e.g., what emerging threats are occurring in media other than e-mail including spam over instant messaging, etc.?); putting consumers back in control (how can we empower consumers and businesses in the fight against spam and malware?); and stakeholder best practices.

STATE LAWS REGULATING SPAM

According to the SpamLaws website [http://www.spamlaws.com], 38 states have passed laws regulating spam: Alaska, Arizona, Arkansas, California, Colorado, Connecticut, Delaware, Florida, Georgia, Idaho, Illinois, Indiana, Iowa, Kansas, Louisiana, Maine, Maryland, Michigan, Minnesota, Missouri, Nevada, New Mexico, North Carolina, North Dakota, Ohio, Oklahoma, Oregon, Pennsylvania, Rhode Island, South Dakota, Tennessee, Texas, Utah, Virginia, Washington, West Virginia, Wisconsin, and Wyoming. The specifics of each law varies. Summaries of and links to each law are provided on that website.[48]

The CAN-SPAM Act preempts state spam laws, but not other state laws that are not specific to electronic mail, such as trespass, contract, or tort law, or other state laws to the extent they relate to fraud or computer crime. California passed an anti-spam law that would have become effective January 1, 2004 and was considered relatively strict. It required opt-in for UCE unless there was a prior business relationship, in which case, opt-out is required. The anticipated implementation of that California law is often cited as one of the factors that stimulated Congress to complete action on a less restrictive, preemptive federal law before the end of 2003.[49]

A number of lawsuits have been filed under the state laws. Two notable cases involve the Maryland and Virginia laws. In December 2004, a Maryland judge ruled that Maryland's anti-spam law is unconstitutional, because it seeks to regulate commerce outside of the state.[50] An individual, Eric Menhart, who was a resident of the District of Columbia, but had a business in Maryland whose domain name was "maryland-state-resident.com," filed suit against a New York-based spammer. According to the spamlaws.com website, the Maryland law prohibits sending commercial e-mail that uses a third party's domain name without permission, or that contains false or missing routing information, or with a false or misleading subject line. The law applies, *inter alia*, to e-mail sent from within Maryland, or if the sender knows that the recipient is a Maryland resident. Mr. Menhart reportedly is appealing the ruling.

A lawsuit brought under Virginia's anti-spam law, however, led to a conviction of two North Carolina residents: Jeremy Jaynes, and his sister, Jessica DeGroot. According to the spamlaws.com website, the Virginia law makes it illegal, *inter alia*, to send unsolicited bulk e-mails containing falsified routing information, and allows the court to exercise personal jurisdiction over a nonresident who uses a computer or computer network located in Virginia. The case reportedly is the first felony spam case in the country. According to press accounts, Mr. Jaynes and Ms. DeGroot were convicted of misrepresenting the origin of e-mails that sold software and other products (a third defendant was acquitted). The e-mails went through AOL servers located in Virginia. Ms. DeGroot's conviction was later overturned, and Mr. Jaynes, who was sentenced to nine years in prison, appealed his conviction;[51] his conviction was upheld by a three-judge panel for the Virginia Court of Appeals on September 5, 2006. Jaynes plans to appeal this decision, as well, but Virginia Attorney General Robert McDonnell said in a statement that his office plans to ask the court to revoke bond and order Jaynes to begin serving his sentence.[52]

REFERENCES

[1]   The origin of the term spam for unsolicited commercial e-mail was recounted in *Computerworld*, April 5, 1999, p. 70: "It all started in early Internet chat rooms and interactive fantasy games where someone repeating the same sentence or comment was said to be making a 'spam.' The term referred to a Monty Python's Flying Circus scene in which actors keep saying 'Spam, Spam, Spam and Spam' when reading options from a menu."

[2]   This report does not address junk mail or junk fax. See CRS Report RL32177, *Federal Advertising Law: An Overview*, by Henry Cohen, or CRS Report RS21647, *Facsimile Advertising Rules Under the Junk Fax Prevention Act of 2005*, by Patricia Moloney Figliola, respectively, for information on those topics.

[3]   Pew Internet and American Life Project. Pew Internet Project Data Memo. May 2007. Available at [http://www.pewinternet.org/pdfs/PIP_Spam_May_2007.pdf].

[4]   Pew Internet and American Life Project. Pew Internet Project Data Memo. May 2007. Available at [http://www.pewinternet.org/pdfs/PIP_Spam_May_2007.pdf].

[5]   See CRS Report RL31408, *Internet Privacy: Overview and Pending Legislation*, by Marcia S. Smith, for more on Internet privacy.

[6]   See [http://www.ferris.com].

[7]   Quoted in: "Digits." *Wall Street Journal*, May 22, 2003, p. B3.

[8]   Pew Internet and American Life Project. Pew Internet Project Data Memo. March 2004. Available at [http://www.pewinternet.org/pdfs/PIP_Data_Memo_on_Spam.pdf].

[9]   See [http://www.dmaconsumers.org/emps.html].

[10] "Spam" generally refers to e-mail, rather than other forms of electronic communication. The term "spim," for example, is used for unsolicited advertising via Instant Messaging. "Spit" refers to unsolicited advertising via Voice Over Internet Protocol (VOIP). Unsolicited advertising on wireless devices such as cell phones is called "wireless spam."

[11] See [http://www.ftc.gov/bcp], [http://onguardonline.gov/index.html], and [http://www.ftc.gov/spam/].

[12] See   [http://www.consumersunion.org/pub/core_product_safety/000210.html]. Additional spam information is available from CU online at [http://www.

consumerreports.org/cro/electronics-computers/computers/internet-and-other-services/net-threats-9-07/spam/0709_net_spam.htm?resultPageIndex=1and resultIndex=1 and searchTerm=spam].

[13] The webpage to file a complaint is [https://rn.ftc.gov/pls/dod/wsolcq$.startup? Z_ORG_CODE=PU01].

[14] See [http://ec.europa.eu/justice_home/fsj/privacy/studies/spam_en.htm].

[15] For example, see Mitchener, Brandon. "Europe Blames Weaker U.S. Law for Spam Surge." *Wall Street Journal*, February 3, 2004, p. B1 (via Factiva).

[16] Sophos Reveals Latest "Dirty Dozen" Spam Producing Countries. Press release, April 7, 2005. The other countries on the list are: China (9.7%), France (3.2%), Spain (2.7%), Canada (2.7%), Japan (2.1%), Brazil (2%), United Kingdom (1.6%), Germany (1.2%), Australia (1.2%), and Poland (1.2). [http://www.sophos.com/pressoffice/news/articles/2005/ 04/sa_dirtydozen05.html].

[17] Nine bills were introduced in the 108th Congress prior to passage of the CAN-SPAM Act: H.R. 1933 (Lofgren), H.R. 2214 (Burr-Tauzin-Sensenbrenner), H.R. 2515 (Wilson-Green), S. 877 (Burns-Wyden), S. 1052 (Nelson-FL), and S. 1327 (Corzine) were "opt-out" bills. S. 563 (Dayton) was a "do not e-mail" bill. S. 1231 (Schumer) combined elements of both approaches. S. 1293 (Hatch) created criminal penalties for fraudulent e-mail.

[18] The Senate originally passed S. 877 on October 22, 2003, by a vote of 97-0. As passed at that time, the bill combined elements from several of the Senate bills. The House passed (392-5) an amended version of S. 877 on November 21, 2003, melding provisions from the Senate-passed bill and several House bills. The Senate concurred in the House amendment, with an amendment, on November 25, through unanimous consent. The Senate amendment included several revisions, requiring the House to vote again on the bill. The House agreed with the Senate amendment by unanimous consent on December 8, 2003.

[19] Some spam already contains instructions, usually to send a message to an e-mail address, for how a recipient can opt-out. However, in many cases this is a ruse by the sender to trick a recipient into confirming that the e-mail has reached a valid e-mail address. The sender then sends more spam to that address and/or includes the e-mail address on lists of e-mail addresses that are sold to bulk e-mailers. It is virtually impossible for a recipient to discern whether the proffered opt-out instructions are genuine or duplicitous.

[20] See [http://www.cauce.org/node/57].

[21] See [http://www.europa.eu.int/scadplus/leg/en/lvb/l24120.htm].

[22] Not all EU nations have yet passed such legislation. According to the Associated Press (December 7, 2003, 12:30), the EU asked nine countries (Belgium, Germany, Greece, Finland, France, Luxembourg, the Netherlands, Portugal, and Sweden) to provide within two months an explanation of when they will pass such legislation. AP identified six countries that have taken steps to implement the EU law: Austria, Britain, Denmark, Ireland, Italy, and Spain. Sweden reportedly adopted spam legislation in March 2004.

[23] A survey by the ePrivacy Group found that 74% of consumers want such a list. Bowman, Lisa. "Study: Do-Not-Spam Plan Winning Support," c|net news.com, July 23, 2003, 12:28 PM PT.

[24] Muris, Timothy. The Federal Trade Commission and the Future Development of U.S. Consumer Protection Policy. Remarks to the Aspen Summit, Aspen, CP, August 19, 2003. [http://www.ftc.gov/speeches/muris/030819aspen.htm].

[25] Krim, Jonathan. "Senate Votes 97-0 to Restrict E-Mail Ads; Bill Could Lead to No-Spam Registry." *Washington Post*, October 23, 2003, p. A1 (via Factiva).

[26] U.S. Federal Trade Commission. Statement of Timothy J. Muris Regarding Passage of the Can-Spam Act of 2003. November 21, 2003. [http://www.ftc.gov/opa/2003/11/spamstmt.htm]

[27] The FTC issued a warning to consumers in February 2004 that a website (unsub.us) promoting a National Do Not Email Registry is a sham and might be collecting e-mail addresses to sell to spammers. See [http://www.ftc.gov/opa/2004/02/spamcam.htm].

[28] U.S. Federal Trade Commission. National Do Not Email Registry: A Report to Congress. Washington, FTC, June 2004. A press release, and a link to the report, is available at [http://www.ftc.gov/opa/2004/06/canspam2.htm].

[29] FTC Survey Tests Top E-Tailers' Compliance with Can-spam's Opt-Out Provisions. August 1, 2005. See [http://www.ftc.gov/opa/2005/08/optout.htm].

[30] See [http://www.ftc.gov/opa/2004/04/adultlabel.htm].

[31] FTC Cracks Down on Illegal "X-Rated" Spam. July 20, 2005. [http://www.ftc.gov/opa/2005/07/alrsweep.htm]

[32] FTC. Subject Line Labeling As A Weapon Against Spam: A Report to Congress. June 17, 2005. p. i-ii. [http://www.ftc.gov/reports/canspam05/050616canspamrpt.pdf]

[33] See CRS Report RL31636, *Wireless Privacy and Spam: Issues for Congress*, for more information.

[34] A press release is available at [http://www.ftc.gov/opa/2004/09/bounty.htm], and the report, A CAN-Spam Informant Reward System, is available at [http://www.ftc.gov/reports/rewardsys/040916rewardsysrpt.pdf].

[35] FTC press release, FTC Issues Final Rule Defining What Constitutes a "Commercial Electronic Mail Message," December 16, 2004.

[36] The FTC's notice of proposed rulemaking had a slightly different definition. The final rule emphasizes that, in the final rule, the definition is the same as in the act.

[37] This explanation is offered on p. 11 of the text of the *Federal Register* notice as it appears on the FTC website at [http://www.ftc.gov/opa/2005/01/primarypurp.htm].

[38] "HoodiaLife" and "HoodiaPlus," was supposed to contain hoodia gordonii and cause significant weight loss; the other, called "HGHLife" and "HGHPlus," was supposed to be a "natural human growth hormone enhancer" that would dramatically reverse the aging process.

[39] (1) FTC Announces First Can-Spam Act Cases. [http://www.ftc.gov/opa/2004/04/040429canspam.htm]; (2) Department of Justice Announces Arrests of Detroit-Area Men on Violations of the 'Can-Spam' Act. [http://www.usdoj.gov/opa/pr/2004/April/04_crm_281.htm].

[40] FTC press release, Court Stops Spammers From Circulating Unwanted Sexually-Explicit E-mails, January 11, 2005. [http://www.ftc.gov/opa/2005/01/globalnetsolutions.htm].

[41] Mangalindan, Mylene. "Web Firms File Spam Suit Under New Law." *Wall Street Journal*, March 11, 2004, p. B4 (via Factiva).

[42] Hines, Matt. "Massachusetts Files Suit Under Can-Spam." C|NET News.com, July 2, 2004, 11:54 am PDT.

[43] Bray, Hiawatha. "Spammer to Pay $25,000 Settlement." *Boston Globe*, October 8, 2004, p. D3 (via Factiva).

[44] "ISPs Push to Collect Money from Spammers." *Communications Daily*, February 18, 2005, p. 9.

[45] FTC. Effectiveness and Enforcement of the CAN-SPAM Act: A Report to Congress. December 2005 [http://www.ftc.gov/reports/canspam05/051220canspamrpt.pdf].

[46] A November 2005 FTC report concluded that anti-spam technologies used by ISPs are very effective in preventing spam from reaching recipients. A press release summarizing the report is available at [http://www.ftc.gov/opa/2005/11/spam3.htm].

[47] The Spam Summit webpage is online at [http://www.ftc.gov/bcp/workshops/spamsummit/]. The page includes links to both days' transcripts.

[48] See CRS Report RL31488, *Regulation of Unsolicited Commercial E-Mail*, by Angie A. Welborn, for a brief review of the state laws and challenges to them.

[49] For example, see Glanz, William. "House Oks Measure Aimed at Spammers; Senate Likely to Approve Changes." *Washington Times*, November 22, 2003, p. A1 (via Factiva).

[50] Baker, Chris. "Maryland Spam Law Ruled Illegal." *Washington Times*, December 15, 2004, p. C6 (via Factiva).

[51] Bruilliard, Karin. "Woman's Spam Conviction Thrown Out." *Washington Post*, March 2, 2005, p. E01 (via Factiva).

[52] Rondeaux, Candace. "Anti-Spam Conviction Is Upheld." *Washington Post*, September 6, 2006, p. B03. Online at [http://www.washingtonpost.com/wp-dyn/content/article/2006/09/05/AR2006090501166_pf.html].

In: Telecommunications and Media Issues
Editors: A. N. Moller, C. E. Pletson, pp. 51-62

ISBN: 978-1-60456-294-1
© 2008 Nova Science Publishers, Inc.

Chapter 3

CONSTITUTIONALITY OF APPLYING THE FCC'S INDECENCY RESTRICTION TO CABLE TELEVISION[*]

Henry Cohen

ABSTRACT

Various federal officials have spoken in favor of extending the Federal Communication Commission's indecency restriction, which currently applies to broadcast television and radio, to cable and satellite television. This report examines whether such an extension would violate the First Amendment's guarantee of freedom of speech.

The FCC's indecency restriction was enacted pursuant to a federal statute that, insofar as it was found constitutional, requires the FCC to promulgate regulations to prohibit the broadcast of indecent programming from 6 a.m. to 10 p.m. The FCC has found that, for material to be "indecent," it "must describe or depict sexual or excretory organs or activities," and "must be patently offensive as measured by contemporary community standards for the broadcast medium."

In 1978, in *Pacifica*, the Supreme Court held that, because broadcast radio and television have a "uniquely pervasive presence" and are "uniquely accessible to children," the government may, during certain times of day, prohibit "[p]atently offensive, indecent material" on these media. In 1996, however, a Supreme Court plurality held that, with respect to "how pervasive and intrusive [television] programming is . . . cable and broadcast television differ little, if at all."

Then, in 2000, the Court held that governmental restrictions on speech on cable television are, unlike those on broadcast media, entitled to strict scrutiny. Thus, whereas, in *Pacifica*, the Court upheld a restriction on "indecent" material on broadcast media without applying strict scrutiny, the Court apparently would not uphold a comparable restriction on "indecent" material on cable television unless the restriction served a compelling governmental interest by the least restrictive means.

It seems uncertain whether the Court would find that denying minors access to "indecent" material on cable television would constitute a compelling governmental interest. Assuming that it would, then, whether or not there is a less restrictive means than a 6 a.m.-to-10 p.m. ban by which to deny minors access to "indecent" material on cable

[*] Excerpted from CRS Report RL33170, dated September 18, 2007.

television, it appears that a strong case may be made that applying the FCC's indecency restrictions to cable television would violate the First Amendment. This is because, as the Supreme Court wrote when it struck down the ban on "indecent" material on the Internet, "the Government may not ʻreduc[e] the adult population . . . to . . . only what is fit for children.'" In addition, the Court, in the 2000 case mentioned above, struck down a speech restriction on cable television, in part because "for two-thirds of the day no household in those service areas could receive the programming, whether or not the household or the viewer wanted to do so."

Various federal officials have spoken in favor of extending the Federal Communication Commission's indecency restriction,[1] which currently applies to broadcast television and radio, to cable and satellite television.[2] This report examines whether such an extension would violate the First Amendment's guarantee of freedom of speech.

INTRODUCTION

The FCC's indecency restriction was enacted pursuant to a federal statute that, insofar as it was found constitutional, requires the FCC to promulgate regulations to prohibit the broadcast of indecent programming from 6 a.m. to 10 p.m.[3] The FCC has found that, for material to be "indecent," it "must describe or depict sexual or excretory organs or activities," and "must be patently offensive as measured by contemporary community standards for the broadcast medium."[4]

Another federal statute makes it a crime to utter "any obscene, indecent, or profane language by means of radio communication."[5] This statute has been applied to pictures as well as words and to broadcast television as well as radio.[6] In *Federal Communications Commission v. Pacifica Foundation*, the Supreme Court held that the statute does not violate the First Amendment when enforced during hours when children are likely to be in the audience.[7]

The Court in *Pacifica* explained:

> [O]f all forms of communication, it is broadcasting that has received the most limited First Amendment protection. Thus, although other speakers cannot be licensed except under laws that carefully define and narrow official discretion, a broadcaster may be deprived of his license and his forum if the Commission decides that such an action would serve "the public interest, convenience, and necessity." Similarly, although the First Amendment protects newspaper publishers from being required to print the replies of those whom they criticize, *Miami Herald Publishing Co. v. Tornillo*, 418 U.S. 241, it affords no such protection to broadcasters; on the contrary, they must give free time to the victims of their criticism. *Red Lion Broadcasting Co. v. FCC*, 395 U.S. 367.
>
> The reasons for these distinctions are complex, but two have relevance to the present case. First, the broadcast media have established a uniquely pervasive presence in the lives of all Americans. Patently offensive, indecent material presented over the airwaves confronts the citizen, not only in public, but in the privacy of the home, where the individual's right to be left alone plainly outweighs the First Amendment rights of an intruder. . . . To say that one may avoid further offense by turning off the radio when he hears indecent language is like saying that the remedy for an assault is to run away after the first blow.

Second, broadcasting is uniquely accessible to children, even those too young to read. . . . Bookstores and motion picture theaters, for example, may be prohibited from making indecent material available to children. We held in *Ginsberg v. New York*, 390 U.S. 629, that the government's interest in the "wellbeing of its youth" and in supporting "parents' claim to authority in their own household" justified the regulation of otherwise protected expression. . . .[8]

In sum, the Court held that, because broadcast radio and television have a "uniquely pervasive presence" and are "uniquely accessible to children," the government may, during certain times of day, prohibit "[p]atently offensive, indecent material" on these media, as such material threatens the well-being of minors and their parents' authority in their own household. Since 1978, however, when the Court decided *Pacifica*, cable television has become more pervasive, thereby rendering broadcast media a less "uniquely pervasive presence." In *Denver Area Educational Telecommunications Consortium, Inc. v. Federal Communications Commission*, a Supreme Court plurality held that, with respect to "how pervasive and intrusive [television] programming is . . . cable and broadcast television differ little, if at all."[9]

If cable and broadcast television differ little, if at all, then, one might argue, they should be treated alike with regard to indecency restrictions. But does this mean, if one accepts that argument, that cable should be treated like broadcast or that broadcast should be treated like cable? In other words, should cable be made subject to the FCC's indecency restriction, or should broadcast no longer be subject to them? This report will consider these questions from a constitutional standpoint, not from a policy standpoint.

Although one might argue that the fact that broadcast media is no longer uniquely pervasive should render *Pacifica* invalid, no court has found that to be the case, and the Supreme Court has cited *Pacifica* with approval in recent years.[10] The FCC's indecency restriction, therefore, appears to remain constitutional as applied to broadcast media.[11] But would it be constitutional to apply the indecency restriction to cable?

CONSTITUTIONALITY OF APPLYING THE FCC'S INDECENCY RESTRICTION TO CABLE TELEVISION

In *United States v. Playboy Entertainment Group, Inc.,* the Supreme Court held that a content-based speech restriction on cable television "can stand only if it satisfies strict scrutiny. If a statute regulates speech based on its content, it must be narrowly tailored to promote a compelling Government interest. If a less restrictive alternative would serve the Government's purpose, the legislature must use that alternative. To do otherwise would be to restrict speech without an adequate justification, a course the First Amendment does not permit. . . . It is rare that a regulation restricting speech because of its content will ever be permissible."[12]

The indecency restriction is content-based; therefore, for its application to cable television to be constitutional, it must meet "strict scrutiny," which means that it must promote a compelling governmental interest and be the least restrictive means to do so. This is the same standard that the Supreme Court applies to speech in newspapers, the Internet, and every other medium except broadcast radio and television.[13] The Court does not apply strict scrutiny to broadcast media because, as noted in the above quotation from *Pacifica*, the Court holds that broadcast media have less First Amendment protection than other media.[14] The Court, therefore, did not apply strict scrutiny in *Pacifica*, and the fact that in *Pacifica* it upheld the constitutionality of the indecency restriction as applied to broadcast media does not imply that it would uphold its constitutionality as applied to cable.

Playboy concerned federal restrictions on a type of "indecent" material on cable television: "signal bleed," which refers to images or sounds that come through to non-subscribers, even though cable operators have "used scrambling in the regular course of business, so that only paying customers had access to certain programs."[15] These restrictions, which are found in section 505 of the Communications Decency Act of 1996, require operators of cable channels "primarily dedicated to sexually-oriented programming" to implement more effective scrambling — to fully scramble or otherwise fully block programming so that non-subscribers do not receive it — or to "time channel," which, under an FCC regulation meant to transmit the programming only from 10 p.m. to 6 a.m.[16]

"To comply with the statute," the Court noted, "the majority of cable operators adopted the second, or 'time channeling,' approach. The effect . . . was to eliminate altogether the transmission of the targeted programming outside the safe harbor period [6 a.m. to 10 p.m.] in affected cable service areas. In other words, for two-thirds of the day no household in those service areas could receive the programming, whether or not the household or the viewer wanted to do so."[17] The Court also noted that "[t]he speech in question was not thought by Congress to be so harmful that all channels were subject to restriction. Instead, the statutory disability applies only to channels 'primarily dedicated to sexually-oriented programming.'"[18]

The Court then applied strict scrutiny to section 505. It did not explicitly say that shielding children from sexually oriented signal bleed was a compelling interest, but it would "not discount the possibility that a graphic image could have a negative impact on a young child."[19] This suggests the possibility that, even if shielding young children from seeing graphic images on cable television is a compelling governmental interest, the Court might not find that shielding older children from such images is a compelling governmental interest. In addition, the Court rejected another interest as compelling: "Even upon the assumption that the Government has an interest in substituting itself for informed and empowered parents, its interest is not sufficiently compelling to justify this widespread restriction on speech."[20] In any case, it was not necessary for the Court to make an explicit finding of a compelling governmental interest, because it held the statute unconstitutional for not constituting the least restrictive means to advance any such interest.

The Court noted that there is "a key difference between cable television and the broadcasting media, which is the point on which this case turns: Cable systems have the capacity to block unwanted channels on a household-by-household basis. . . . [T]argeted blocking enables the Government to support parental authority without affecting the First Amendment interests of speakers and willing listeners . . . ."[21] Furthermore, targeted blocking is already required — by section 504 of the Communications Decency Act, which requires cable operators, upon request by a cable service subscriber, to, without charge, fully scramble or otherwise fully block audio and video programming that the subscriber does not wish to receive.[22] "When a plausible, less restrictive alternative is offered to a content-based speech restriction, it is the Government's obligation to prove that the alternative will be ineffective to achieve its goal. The Government has not met that burden here."[23] The Court concluded, therefore, that section 504, with adequate publicity to parents of their rights under it, constituted a less restrictive alternative to section 505.

Applying Strict Scrutiny

We now consider how a court might apply strict scrutiny in determining whether applying the FCC's indecency restriction to cable television would be constitutional. We consider first whether a court would find that applying the restriction to cable television would serve a compelling governmental interest, and then, on the assumption that it would, we consider whether a court would find it the least restrictive means to advance that interest.

Compelling Governmental Interest

When the Court considers the constitutionality of a restriction on speech, it ordinarily — even when the speech that is restricted lacks full First Amendment protection and the Court applies less than strict scrutiny — requires the government to "demonstrate that the recited harms are real, not merely conjectural, and that the regulation will in fact alleviate these harms in a direct and material way."[24] With respect to restrictions designed to deny minors access to sexually explicit material, by contrast, the courts appear to assume, without requiring evidence, that such material is harmful to minors, or to consider it "obscene as to minors," even if it is not obscene as to adults,[25] and therefore not entitled to First Amendment protection with respect to minors, whether it is harmful to them or not.[26] In *Pacifica*, as quoted above, the Court implied that making "indecent" material unavailable to children serves their "well-being."[27]

This is not to say with certainty that the Supreme Court would find a compelling governmental interest in denying minors access to "indecent" material. It might, for example, distinguish among different types of "indecent" material, and, even if it found a compelling governmental interest in denying minors access to sexually explicit material, it might find otherwise with respect to four-letter words, in light of the fact that minors generally hear such words elsewhere than on cable television, and in light of the fact that such words may be used as adjectives or expletives, arguably with no sexual or excretory connotation.[28] The Court might also distinguish among minors of different ages, even with respect to access to sexually explicit material. As noted above, in *Playboy* the Court seemed to leave open the possibility that it might not find a compelling governmental interest in shielding older children from sexually oriented material. In addition, when the Court struck down the portion of the Communications Decency Act of 1996 that prohibited "indecent" material on the Internet, the Court would "neither accept nor reject the Government's submission that the First Amendment does not forbid a blanket prohibition on all 'indecent' and 'patently offensive' messages communicated to a 17-year-old — no matter how much value the message may have and regardless of parental approval. It is at least clear that the strength of the Government's interest in protecting minors is not equally strong throughout the coverage of this broad statute."[29]

The Supreme Court has cited another governmental interest that might be asserted to justify applying the FCC's indecency restriction to cable, but the Court has not stated whether it is "compelling": it is the interest "in supporting 'parents' claim to authority in their own household.'"[30] A dissenting judge has argued that "a law that effectively *bans* all indecent programming . . . does not facilitate parental supervision. In my view, my right as a parent has been preempted, not facilitated, if I am told that certain programming will be banned from my . . . television. Congress cannot take away my right to decide what my children watch, absent some showing that my children are in fact at risk of harm from exposure to indecent programming."[31]

Perhaps, however, the Supreme Court would take the approach it did in *Playboy* and focus on the second aspect of strict scrutiny: whether the FCC's indecency restriction is the least restrictive means available to advance the government's interest.

Least Restrictive Means

Assuming for the sake of argument that applying the FCC's indecency restriction to cable television would serve a compelling governmental interest, is there a less restrictive means by which that interest could be served? If so, then applying the FCC's indecency restriction to cable television would be unconstitutional. The Court in *Playboy*, as quoted above, noted that there is "a key difference between cable television and the broadcasting media, which is the point on which this case turns: Cable systems have the capacity to block unwanted channels on a household-by-household basis. . . . [T]argeted blocking enables the Government to support parental authority without affecting the First Amendment interests of speakers and willing listeners . . . ."[32]

The targeted blocking, however, that the Court in *Playboy* found to be a less restrictive means to keep signal bleed from viewers who object to it, would not seem as feasible to keep "indecent" material from viewers who object to it. In the case of signal bleed, a viewer could request blocking of channels that he knows to present pornography. By contrast, in order for targeted blocking to keep "indecent" programming from viewers who object to it, viewers would have to know what channels would ever, between 6 a.m. and 10 p.m., allow on any program the utterance of a four-letter word or the exposure of a woman's breast, among other things.[33] For viewers to know this would seem to entail that every channel be required to state whether it will refrain from transmitting "indecent" material on all future programming, and, if a channel stated that it would, to be bound by that statement. This requirement would appear to burden freedom of speech to the extent that it might well violate the First Amendment.[34] It might even be viewed as a prior restraint, and prior restraints are almost always unconstitutional.[35]

To the extent that technology allows viewers to block particular programs as opposed to entire channels, the same First Amendment difficulties would apparentlyarise. For such blocking to be effective, producers who did not wish to be blocked for transmitting "indecent" programming would have to agree to refrain from ever allowing the utterance of a four-letter word on a program, even if the program ordinarily contained nothing deemed "indecent."

Thus, there may be no less restrictive means that would be constitutional to keep "indecent" material off cable television during certain hours than to apply the FCC's indecency restrictions.[36] This, however, would not necessarily mean that to apply them to cable television would be constitutional. Two federal courts of appeals have written that "[t]he State may not regulate at all if it turns out that even the least restrictive means of regulation is still unreasonable when its limitations on freedom of speech are balanced against the benefits gained from those limitations."[37] The more recent of these cases affirmed a preliminary injunction against the enforcement of the Child Online Protection Act, which banned material that is "harmful to minors" from the Internet.[38] The older case upheld FCC regulations that implemented a statute that restricted minors' access to obscene "dial-a-porn" services.

It appears that a strong case may be made that applying the FCC's indecency restriction to cable television would be "unreasonable" under the above court of appeals' formulation. This is because, as the Supreme Court wrote when it struck down the ban on "indecent" material on the Internet, "the Government may not 'reduc[e] the adult population . . . to . . . only what is fit for children.' '[R]egardless of the strength of the government's interest' in protecting children, '[t]he level of discourse reaching a mailbox simply cannot be limited to that which would be suitable for a sandbox.'"[39]

One might reply that to apply the FCC's indecency restriction to cable would limit the adult population's discourse only from 6 a.m. to 10 p.m. But the fact the Supreme Court in *Pacifica* upheld such a limitation on broadcast media does not mean that it would uphold it on cable television. In *Pacifica*, as noted, the Court did not apply strict scrutiny. In *Playboy*, where the Court did apply strict scrutiny to a speech restriction on cable television, it held the speech restriction unconstitutional, in part because, as quoted above, "for two-thirds of the day no household in those service areas could receive the programming, whether or not the household or the viewer wanted to do so."[40] In addition, the Court struck down the ban in the Communications Decency Act of 1996 on "indecent" material on the Internet, notwithstanding that such material is available to adults in other media.[41] It seems clear that governmental restrictions of fully protected speech, including "indecent" material on cable television, are unconstitutional unless they pass strict scrutiny, even if they do not close all outlets for such speech.

Summary and Conclusion

In 1978, in *Pacifica*, the Supreme Court held that, because broadcast radio and television have a "uniquely pervasive presence" and are "uniquely accessible to children," the government may, during certain times of day, prohibit "[p]atently offensive, indecent material" on these media. In 1996, however, in *Denver Area Consortium,* a Supreme Court plurality held that, with respect to "how pervasive and intrusive [television] programming is . . . cable and broadcast television differ little, if at all."

The fact that a plurality of the Court views cable and broadcast television as differing little with respect to their pervasiveness and intrusiveness might suggest that the Court would apply the First Amendment to both media in the same way. The Court, however, continues to cite *Pacifica* with approval, but, in *Playboy*, it held that governmental restrictions on cable television are, unlike those on broadcast media, entitled to strict scrutiny. Thus, whereas, in *Pacifica*, the Court upheld a restriction on "indecent" material on broadcast media without applying strict scrutiny, the Court apparently would not uphold a comparable restriction on "indecent" material on cable television unless the restriction served a compelling governmental interest by the least restrictive means.

It seems uncertain whether the Court would find that denying minors access to "indecent" material on cable television would constitute a compelling governmental interest. Although the Court has held that denying minors access to sexually explicit material constitutes a compelling governmental interest, not all "indecent" material is sexually explicit. In addition, the Court has suggested that it may not view minors of all ages identically for First Amendment purposes.

Assuming for the sake of argument that the Court would find a compelling governmental interest in denying minors access to "indecent" material on cable television, there does not appear to be a less restrictive means than the FCC's restrictions to advance this interest, other than banning "indecent" material for fewer hours per day. The lack of a less restrictive means, however, would not necessarily mean that to apply the FCC's restrictions to cable television would be constitutional. This is because, as two federal courts of appeals have written, "[t]he State may not regulate at all if it turns out that even the least restrictive means of regulation is still unreasonable when its limitations on freedom of speech are balanced against the benefits gained from those limitations." The Supreme Court has not spoken on this proposition, however.

It appears that a strong case may be made that applying the FCC's indecency restriction to cable television would be "unreasonable" under this formulation. This is because, as the Supreme Court wrote when it struck down the ban on "indecent" material in the Internet, "the Government may not 'reduc[e] the adult population . . . to . . . only what is fit for children.'" In *Playboy*, the Court, applying strict scrutiny, struck down a speech restriction on cable television, in part because "for two-thirds of the day no household in those service areas could receive the programming, whether or not the household or the viewer wanted to do so." Thus, it appears likely that a court would find that to apply the FCC's indecency restriction to cable television would be unconstitutional.

REFERENCES

[1]   47 C.F.R. § 73.3999(b).
[2]   This report hereinafter does not refer to satellite television, but its references to cable television to may read to include satellite television. For references to support for applying the indecency restriction to cable television, see, e.g., Paul K. McMasters, *Inside the First Amendment: Surrendering our choices to a sense of decency*, Gannet News Service (April 11, 2005); Drew Clark, *Lawmakers May Only Be Partially Pleased by Cable*, National Journal's Technology Daily (April 4, 2005).
[3]   47 U.S.C. § 303 note; Action for Children's Television v. Federal Communications Commission, 58 F.3d 654 (D.C. Cir. 1995) (en banc), *cert. denied*, 516 U.S. 1043 (1996) (striking down the statute insofar as it banned "indecent programming" on non-public broadcast stations from 10 p.m. to midnight).
[4]   In the Matter of Industry Guidance on the Commission's Case Law Interpreting 18 U.S.C. § 1464 and Enforcement Policies Regarding Broadcast Indecency, File No. EB-00-IH-0089 (April 6, 2001). [http://www.fcc.gov/eb/Orders/2001/fcc01090.html].
[5]   18 U.S.C. § 1464.
[6]   See, e.g., Complaints Against Various Television Licensees Concerning Their February 1, 2004, Broadcast of the Super Bowl XXXVIII Halftime Show, File No. EB-04-IH-0011 (September 22, 2004).

[7]   438 U.S. 726 (1978) ("It is appropriate, in conclusion, to emphasize the narrowness of our holding. . . . The time of day was emphasized by the Commission. The content of the program in which the language is used will also affect the composition of the audience." *Id.* at 750). A federal court of appeals later declared a 24-hour-a-day ban unconstitutional. *Action for Children's Television v. Federal Communications Commission*, 932 F.2d 1504 (D.C. Cir. 1991), *cert denied*, 503 U.S. 913 (1992).

[8]   *Id.* at 748-749.

[9]   518 U.S. 727, 748 (1996). The plurality, quoting words from *Pacifica* that appear in the indented quotation above, added that cable television "is as 'accessible to children' as over-the-air broadcasting, if not more so," has also "established a uniquely pervasive presence in the lives of all Americans," and can also "'confron[t] the citizen' in 'the privacy of the home,' . . . with little or no prior warning." *Id.* at 744-745. Justice Souter concurred that "today's plurality opinion rightly observes that the characteristics of broadcast radio that rendered indecency particularly threatening in *Pacifica*, that is, its intrusion into the house and accessibility to children, are also present in the case of cable television. . . ." *Id.* at 776.

[10] Reno v. American Civil Liberties Union, 521 U.S. 844, 868 (1997); United States v. Playboy Entertainment Group, Inc., 529 U.S. 803, 813 (2000); Ashcroft v. Free Speech Coalition, 535 U.S. 234, 245 (2002).

[11] This is not to say that every application of them by the FCC is necessarily constitutional. The FCC's application of the restrictions to Bono's single use of a four-letter word as a modifier, and to Janet Jackson's "wardrobe malfunction" during a Superbowl halftime show, are being challenged; *see* CRS Report RL32222, *Regulation of Broadcast Indecency: Background and Legal Analysis*, by Henry Cohen and Kathleen Ann Ruane.

[12] *Playboy, supra* note 10, at 813, 818 (citations omitted).

[13] The Court in *Playboy* wrote, however, "Cable television, like broadcast media, presents unique problems, which inform our assessment of the interests at stake, and which may justify restrictions that would be unacceptable in other contexts." *Id.* Nevertheless, because the Court applied strict scrutiny in *Playboy*, this comment apparently means not that the Court would apply less than strict scrutiny to restrictions on cable television, but that applying strict scrutiny to restrictions on cable television might involve considerations not present when applying it to restrictions on other media.

[14] The lower level of First Amendment protection for broadcast media dates back to *Red Lion Broadcasting Co. v. Federal Communications Commission*, 395 U.S. 367, 369 (1969), in which the Supreme Court upheld the FCC's "fairness doctrine," which "imposed on radio and television broadcasters the requirement that discussion of public issues be presented on broadcast stations, and that each side of those issues must be given fair coverage." The reason that the Court upheld the imposition of the fairness doctrine on broadcast media, though it would not uphold its imposition on print media, is that "[w]here there are substantially more individuals who want to broadcast than there are frequencies to allocate, it is idle to posit an unabridgeable First Amendment right to broadcast comparable to the right of every individual to speak, write, or publish." *Id.* at 388. In *Turner Broadcasting System, Inc. v. Federal Communications Commission*, 512 U.S. 622, 639 (1994), the Court held that this "spectrum scarcity" problem does not apply to cable television. In *Denver Area, supra* note 9, 518 U.S. at

748, the Court noted that spectrum scarcity, in any event, "has little to do with a case that involves the effects of television viewing on children. Those effects are the result of . . . how pervasive and intrusive that programming is."

[15] *Id.* at 806.

[16] 47 U.S.C. § 561.

[17] *Playboy, supra* note 10, at 806-807.

[18] *Id.* at 812.

[19] *Id.* at 826.

[20] *Id.* at 825.

[21] *Id.* at 815.

[22] 47 U.S.C. § 560.

[23] *Playboy, supra* note 10, 529 U.S. at 816.

[24] *Turner Broadcasting, supra* note 14, 512 U.S. at 664 (incidental restriction on speech). *See also,* Edenfield v. Fane, 507 U.S. 761, 770-771 (1993) (restriction on commercial speech); Nixon v. Shrink Missouri Government PAC, 528 U.S. 377, 392 (2000) (restriction on campaign contributions). In all three of these cases, the government had restricted less-than-fully protected speech, so the Court did not apply strict scrutiny. Because "indecent" material is generally entitled to full First Amendment protection (except on broadcast media), one might expect that the Court, in determining the constitutionality of applying the FCC's indecency restriction to cable television, would be all the more likely to require the government to demonstrate that harms it recites are real and that the indecency restriction would alleviate these harms in a direct and material way. But see the next sentence in the text.

[25] Material that is obscene as to adults is not entitled to First Amendment protection, whether or not it is harmful to adults. *Miller v. California,* 413 U.S. 15 (1973); *see* CRS Report 95-804, *Obscenity and Indecency: Constitutional Principles and Federal Statutes,* by Henry Cohen.

[26] Interactive Digital Software Association v. St. Louis County, Missouri, 329 F.3d 954, 959 (8th Cir. 2003). The Supreme Court has "recognized that there is a compelling interest in protecting the physical and psychological well-being of minors. This interest extends to shielding minors from the influence of literature that is not obscene by adult standards." *Sable Communications of California v. Federal Communications Commission,* 492 U.S. 115, 126 (1989). The Court has also upheld a state law banning the distribution to minors of "so-called 'girlie' magazines," even as it acknowledged that "[i]t is very doubtful that this finding [that such magazines are "a basic factor in impairing the ethical and moral development of our youth"] expresses an accepted scientific fact." *Ginsberg v. New York,* 390 U.S. 629, 631, 641. "To sustain state power to exclude [such material from minors]," the Court wrote, "requires only that we be able to say that it was not irrational for the legislature to find that exposure to material condemned by the statute is harmful to minors." *Id.* at 641. *Ginsberg* thus "invokes the much less exacting 'rational basis' standard of review," rather than strict scrutiny. *Interactive Digital Software Association, supra,* 329 F.3d at 959. In addition, before *Playboy* reached the Supreme Court, a federal district court wrote: We are troubled by the absence of evidence of harm presented both before Congress and before us that the viewing of signal bleed of sexually explicit programming causes harm to children and that the avoidance of this harm can be recognized as a compelling State interest. We

recognize that the Supreme Court's jurisprudence does not require empirical evidence. Only some minimal amount of evidence is required when sexually explicit programming and children are involved. *Playboy Entertainment Group, Inc. v. United States*, 30 F. Supp.2d 702, 716 (D. Del. 1998), *aff'd*, 529 U.S. 803 (2000). The district court therefore found that the statute served a compelling governmental interest, though it held it unconstitutional because it found that the statute did not constitute the least restrictive means to advance the interest. As noted in the text above, the Supreme Court affirmed on the same ground. In another case, a federal court of appeals, upholding the FCC's indecency restrictions as applied to broadcast media, noted "that the Supreme Court has recognized that the Government's interest in protecting children extends beyond shielding them from physical and psychological harm. The statute that the Court found constitutional in *Ginsberg* sought to protect children from exposure to materials that would 'impair[ ] [their] *ethical and moral* development. . . . Congress does not need the testimony of psychiatrists and social scientists in order to take note of the coarsening of impressionable minds that can result from a persistent exposure to sexually explicit material. . . .'" *Action for Children's Television v. Federal Communications Commission, supra* note 3, 58 F.3d at 662 (brackets and italics supplied by the court of appeals). A dissenting judge in the case noted that, "[t]here is not one iota of evidence in the record . . . to support the claim that exposure to indecency is harmful — indeed, the nature of the alleged 'harm' is never explained." *Id.* at 671 (Edwards, C.J., dissenting).

[27]   *See*, note 8, *supra*.

[28]   The FCC ruled that "given the core meaning of the 'F-Word,' any use of that word, or a variation, in any context, inherently has a sexual connotation. . . ." In the Matter of Complaints Against Various Broadcast Licensees Regarding Their Airing of the "Golden Globe Awards" Program, File No. EB-03-IH-0110 (March 18, 2004). The U.S. Court of Appeals for the Second Circuit, however, found that this view "defies any commonsense understanding of these words, which, as the general public knows, are often used in everyday conversation without any 'sexual or excretory' meaning." Fox Television Stations, Inc. v. Federal Communications Commission, 489 F.3d 444, 459 (2d Cir. 2007).

[29]   Reno v. American Civil Liberties Union, *supra* note 10, 521 U.S. at 878.

[30]   *Pacifica, supra* note 7, 438 U.S. at 749, quoting *Ginsberg, supra* note 26, 390 U.S. at 639. *See also, Playboy, supra* note 10, 529 U.S. at 815.

[31]   Action for Children's Television v. Federal Communications Commission, *supra* note 3, 58 F.3d at 670 (emphasis in original) (Edwards, C.J., dissenting).

[32]   *Playboy, supra* note 10, 529 U.S. at 815.

[33]   One commentator, referring to the present CRS report, writes, "The Congressional Research Service misses an important distinction. In *Playboy*, the Court was dealing with a new federal statute that would lead to a 6:00 - 10:00 p.m. ban on indecent material being aired on *primarily sexually-explicit channels*. By its very definition, this was *expected* indecency, and the lockbox and other technological solutions . . . would actually work as intended. . . . For a pay-per-view cable channel that peddles exclusively in smut, blocking technologies are highly effective remedies. The problem is that blocks are totally ineffective when dealing with *unexpected* indecency. Protecting the children, while not compelling enough to warrant a draconian block on

all sexual programming from adults who explicitly request it, is likely still compelling enough to warrant the prohibition of indecency on general cable channels." Matthew S. Schwartz, *A Decent Proposal: The Constitutionality of Indecency Regulation on Cable and Direct Broadcast Satellite Services*, 13 Richmond Journal of Law and Technology 17, 27 (2007). In fact, the sentence of the text above that ends with the present footnote — and the sentence has not been revised since the first edition of this report — makes the distinction to which the commentator refers. This report, however, explains below why this distinction may be insufficient to make the extension of the FCC's indecency rule to cable television constitutional.

[34] "The distinction between laws burdening and laws banning speech is but a matter of degree. The Government's content-based burdens must satisfy the same rigorous scrutiny as it content-based bans." *Playboy, supra* note 10, at 812.

[35] See CRS Report 95-815, Freedom of Speech and Press: Exceptions to the First Amendment, by Henry Cohen.

[36] The government could, of course, reduce the numbers of hours per day that it bans "indecent" material, though the Supreme Court might be less likely to find a compelling governmental interest in a shorter ban, because a shorter ban would be less effective in denying minors access to "indecent" material.

[37] American Civil Liberties Union v. Reno, 217 F.3d 162, 179 (3d Cir. 2000), *vacated and remanded on other grounds sub nom.*, Ashcroft v. American Civil Liberties Union, 535 U.S. 564 (2002), quoting Carlin Communications, Inc. v. Federal Communications Commission, 837 F.2d 546, 555 (2d Cir. 1988), *cert. denied*, 488 U.S. 924 (1988).

[38] In *Reno v. American Civil Liberties Union, supra* note 10, 521 U.S. 844 (1997), the Supreme Court struck down the Communications Decency Act of 1996, which banned "indecent" material on the Internet, and, in *Ashcroft v. American Civil Liberties Union*, 542 U.S. 656 (2004) (a second Supreme Court decision in *American Civil Liberties Union v. Reno, supra* note 37), the Court upheld a preliminary injunction against enforcement of the Child Online Protection Act, which banned "harmful to minors" material on the Internet. In both cases, the Court applied strict scrutiny and suggested that filtering software might be a less restrictive alternative. 521 U.S. at 844; 542 U.S. at 668. This alternative, of course, would not be an option with respect to cable television.

[39] *Reno, supra* note 10, 521 U.S. at 875 (citations omitted).

[40] *Playboy, supra* note 10, at 806-807.

[41] *Reno, supra* note 10.

In: Telecommunications and Media Issues
Editors: A. N. Moller, C. E. Pletson, pp. 63-73

ISBN: 978-1-60456-294-1
© 2008 Nova Science Publishers, Inc.

Chapter 4

THE TRANSITION TO DIGITAL TELEVISION: IS AMERICA READY?[*]

Lennard G. Kruger

ABSTRACT

The Deficit Reduction Act of 2005 (P.L. 109-171) directs that on February 18, 2009, over-the-air television broadcasts — which are currently provided by television stations in both analog and digital formats — will become digital only. Digital television (DTV) technology allows a broadcaster to offer a single program stream of high definition television (HDTV), or alternatively, multiple video program streams (multicasts). Households with over-the-air analog-only televisions will *no longer be able to receive television service* unless they either: (1) buy a digital-to-analog converter box to hook up to their analog television set; (2) acquire a digital television or an analog television equipped with a digital tuner; or (3) subscribe to cable, satellite, or telephone company television services, which will likely provide for the conversion of digital signals to their analog customers.

Households using analog televisions for viewing over-the-air television broadcasts are likely to be most affected by the digital transition. Of particular concern to many policymakers are low-income, elderly, disabled, non-English speaking, and minority populations. Many of these groups tend to rely more on over-the-air television, and are thus more likely impacted by the digital transition.

The Deficit Reduction Act of 2005 established a digital-to-analog converter box program — administered by the National Telecommunications and Information Administration (NTIA) of the Department of Commerce — that will partially subsidize consumer purchases of converter boxes. NTIA will provide up to two forty-dollar coupons to requesting U.S. households. The coupons are to be issued between January 1, 2008, and March 31, 2009, and must be used within three months after issuance towards the purchase of a stand-alone device used solely for digital-to-analog conversion.

The preeminent issue for Congress is ensuring that American households are prepared for the February 17, 2009 DTV transition deadline, thereby minimizing a scenario whereby television sets across the nation "go dark." Specifically, Congress is actively overseeing the activities of federal agencies responsible for the digital transition

---

[*] Excerpted from CRS Report RL34165, dated October 9, 2007.

— principally the Federal Communications Commission (FCC) and the NTIA — while assessing whether additional federal efforts are necessary, particularly with respect to public education and outreach. The Congress is also monitoring the extent to which private sector stakeholders take appropriate and sufficient steps to educate the public and ensure that all Americans are prepared for the digital transition. DTV- related bills, which address public education (H.R. 608, H.R. 2566, H.R. 2917, and S. 2125), have been introduced into the 110th Congress. At issue is whether the federal government's current programs and reliance on private sector stakeholders will lead to a successful digital transition with a minimum amount of disruption to American TV households or, alternatively, whether additional legislative measures are warranted.

This report will be updated as events warrant.

INTRODUCTION

Under current law, after February 17, 2009, households with over-the-air analog-only televisions will *no longer be able to receive television service* unless they either: (1) buy a digital-to-analog converter box to hook up to their analog television set; (2) acquire a digital television or an analog television equipped with a digital tuner;[1] or (3) subscribe to cable, satellite, or telephone company television services, which will likely provide for the conversion of digital signals to their analog customers. The Deficit Reduction Act of 2005 (P.L. 109-171) directs that on February 18, 2009, over-the-air television broadcasts — which are currently provided by television stations in both analog and digital formats — will become digital only.[2] Analog broadcast television signals, which have been broadcast for over 60 years, will cease, and television stations will broadcast exclusively digital signals over channels 2 through 51.

The preeminent issue for Congress is ensuring that American households are prepared for the transition, thereby minimizing a scenario whereby television sets across the nation "go dark" on February 18, 2009. Specifically, Congress is actively overseeing the activities of federal agencies responsible for the digital transition —principally the Federal Communications Commission (FCC) and the National Telecommunications and Information Administration (NTIA) — while assessing whether additional federal efforts are necessary. The Congress is also monitoring the extent to which private sector stakeholders take appropriate and sufficient steps to educate the public and ensure that all Americans are prepared for the digital transition.

WHAT IS DIGITAL TELEVISION?

Digital television (DTV) is a new television service representing the most significant development in television technology since the advent of color television. DTV can provide movie-quality pictures and sound far superior to traditional analog television. Digital television technology allows a broadcaster to offer a single program stream of high definition television (HDTV) or, alternatively, multiple video program streams ("multicasts") of standard or enhanced definition television, which provide a lesser quality picture than HDTV, but a generally better picture than analog television. DTV technology also makes possible an interactive capability, such as "pay-per-view" service over-the-air.

In order to receive and view digital high-definition television service, consumers must have a digital television set equipped with a digital tuner capable of receiving the digital signal that is provided either over-the-air (in which case an antenna is required) or via cable or satellite television systems. Additionally, consumers can view high definition programs with a digital TV attached to a high definition DVD player (i.e., the HD-DVD or Blu-Ray Disc player).

WHY IS THE NATION TRANSITIONING TO DIGITAL TELEVISION?

One of the key drivers behind the digital transition is reclaiming a portion of the analog spectrum (broadcast channels 52 through 69, also known as the 700 MHZ band) currently occupied by television broadcasters. Digital television uses radio frequency spectrum more efficiently than traditional analog television, thereby "freeing up" bandwidth. The goal of the FCC and Congress has been to complete the transition to DTV as quickly as is possible and feasible, so that analog spectrum can be reclaimed and subsequently reallocated for other purposes. Some of the analog spectrum will be auctioned for commercial wireless services (including wireless broadband), and some will be used for new public safety communications services. Additionally, it is mandated that some of the revenue raised in the spectrum auction will be returned to the U.S. Treasury, thereby contributing toward federal deficit reduction. For more information on the auction and use of the analog television spectrum, see CRS Report RS22218, *Spectrum Use and the Transition to Digital TV*, by Linda K. Moore.

Another rationale often cited for the digital transition is that — aside from offering a superior television viewing experience to consumers — DTV will give over-the-air broadcasters the capability to offer more channels of programming (via multicasting, if they so choose) as well as the ability to offer similar digitally-based services (such as pay-per-view or other interactive services) offered by cable and satellite television providers.

WHO IS LIKELY TO BE MOST AFFECTED BY THE TRANSITION?

Households using analog televisions for viewing over-the-air television broadcasts are likely to be most affected by the digital transition. Estimates vary over the number of analog TV sets and households affected. For example, the National Association of Broadcasters (NAB) has estimated that there are 69 million analog television sets that will be potentially impacted by the digital transition,[3] consisting of 19.6 million households (17% of all households) relying exclusively on over-the-air analog television sets and an additional 14.7 million cable and satellite households receiving some over-the-air programming on analog sets.[4] On the other hand, the Consumer Electronics Association (CEA) has estimated that 36.5 million analog televisions (comprising 13.5 million households) will require converter boxes; according to CEA, an additional 30 million analog sets are used for non-broadcast purposes such as playing video games or watching DVDs, and will therefore likely not require converter boxes.[5]

Of particular concern to many policymakers are low-income, elderly, disabled, non-English speaking, and minority populations. Many of these groups tend to rely more on over-the-air television, and are thus more likely to be impacted by the digital transition. A survey commissioned by the Association of Public Television Stations (APTS) indicated that Americans aged 65 and older are consistently more likely to receive television signals via an over-the-air antenna than are Americans under 65. The survey found that during the first quarter of 2007, 24% of households with Americans 65 and older received their TV programming over-the-air, while only 19% of younger households were over-the-air. The study also found that of Americans aged 65 and older who rely solely on over-the-air connections to television programming, only 17% own a digital TV.[6]

Similarly, a 2005 Government Accountability Office (GAO) survey found that over-the-air households are more likely to have lower incomes than cable or satellite households. Specifically, GAO found that approximately 48% of exclusive over-the-air viewers have household incomes less than $30,000, versus 6% with household incomes over $100,000. GAO also found that nonwhite and Hispanic households are more likely to rely on over-the-air television, with over 23%of non-white households relying on over-the-air television compared to less than 16% of white households, and about 28%of Hispanic households relying on over-the-air television compared to about 17% of non-Hispanic households.[7]

HOW WILL THE DIGITAL TRANSITION AFFECT CABLE AND SATELLITE HOUSEHOLDS?

Multichannel video programming distributor (MVPD) households — consisting of households receiving cable, satellite, or telephone company television services —constitute approximately 85% of all U.S. television households. Many of these households will likely continue to use analog televisions after the transition. For those customers, it is expected that providers will handle the digital-to-analog conversion, either at the "head end" by providing downconverted analog signals, or at the customer premises via a set top box provided by the cable or satellite company. At the same time, many cable and satellite households also have spare televisions relying on over-the-air broadcasts. These stand-alone over-the-air analog televisions will no longer function unless they are equipped with a converter box.

On September 11, 2007, the FCC adopted rules intended to ensure that cable customers continue to receive local TV stations after the transition. Specifically, the FCC will require cable operators to comply with a "viewability requirement" by choosing to either 1) carry the signal in analog as well as digital formats (dual carriage), or 2) carry the signal in a digital only format, provided that all subscribers have set-top boxes which will enable them to view digital broadcasts on their analog TVs. The viewability requirement extends to February 2012, at which time the FCC will reassess the need for the requirement. Small cable companies — which had sought an exemption — may request a waiver of the viewability requirement.

THE DIGITAL-TO-ANALOG CONVERTER BOX PROGRAM

After February 17, 2009, analog-only televisions will no longer be able to receive over-the-air broadcast signals, unless those televisions are equipped with a digital-to-analog converter box. A separate converter box — expected to be available for $50 to $70 — will be required for each analog over-the-air television set. Converter boxes will not only enable all analog televisions to function, they should also provide better reception, additional features such as closed captioning and parental controls, and allow the viewing of multicasted channels. However, a converter box hooked up to an analog TV will not enable the viewer to watch a broadcast in the high-definition format.

The 109th Congress acted to establish a digital-to-analog converter box program that will partially subsidize consumer purchases of converter boxes. Title III of the Deficit Reduction Act of 2005 (P.L. 109-171) directs the National Telecommunications and Information Administration (NTIA) of the Department of Commerce to provide up to two forty-dollar coupons to requesting U.S. households. According to the statute, the coupons are to be issued between January 1, 2008, and March 31, 2009, and must be used within three months after issuance towards the purchase of a stand-alone device used solely for digital-to-analog conversion.

The converter box program will be funded by receipts from the auction of the analog television spectrum. P.L. 109-171 designates $990 million for the converter box program, including up to $100 million for administrative costs (of which no more than $5 million can be used for consumer education). In the event that NTIA notifies Congress that additional funding is needed, the total may be raised up to $1.5 billion, including up to $160 million for administrative costs.

On March 12, 2007, NTIA released its final rule implementing the converter box program.[8] The final rule states that starting on January 1, 2008, for the initial $990 million program (the "Initial Period"), up to two forty-dollar coupons will be available to any and all requesting U.S. households to be used towards the purchase of up to two digital-to-analog converter boxes. Coupons mailed to consumers will be accompanied by information listing converter box models and local (and online) retailers certified to participate in the converter box coupon program. In the event that NTIA determines that the additional $510 million is needed, only exclusively over-the-air households will be eligible for coupons during this "Contingent Period."

During the "Contingent Period," households will be required to self-certify that they are exclusively over-the-air and do not subscribe to cable, satellite, or other pay television services. Cable and satellite households that contain extra over-the-air televisions will be eligible for coupons during the "Initial Period" of the program (the first $990 million), but will not be eligible for coupons if there is a second phase ("Contingent Period") of the program (the additional $510 million).

The rule also sets forth procedures and requirements for manufacturers and retailers who wish to participate in the converter box program. Participation in the converter box program is voluntary. Manufacturers must submit test results and sample converter boxes to NTIA for approval. Approved devices must meet prescribed technical specifications that are intended to ensure an affordable state-of-the-art converter box. Additional permitted features include a smart antenna interface connector and program guide. Features that would disqualify a converter box from being covered by the coupon program include video recording, playback capability, or other capabilities that allow more than simply converting a digital over-the-air signal.[9]

Meanwhile, retailers must receive a certification from NTIA in order to participate in the converter box coupon program. Certified retailers must agree to have systems in place capable of processing coupons electronically for redemption and payment, track every transaction and provide reports to NTIA, train employees on the purpose and operation of the coupon program with NTIA-provided training materials, use commercially reasonable methods to order and manage inventory, and assist NTIA in minimizing incidents of waste, fraud, and abuse, including reporting suspicious patterns of customer behavior. Retailers are not responsible for verifying household eligibility.[10]

On August 15, 2007, NTIA announced it had entered into a contract with IBM to run the Digital-to-Analog Converter Box Coupon program. The total award is $119,968,468, which breaks down to $84,990,343 for the initial period and $34,978,125 for the contingent period. The contract performance began immediately and is to close out on September 30, 2009. The IBM-led team will provide services in three areas: consumer education, coupon distribution to consumers and retail store participation, and financial processing to reimburse retailers, to maintain records, and to prevent fraud, waste, and abuse.

Status of DTV Public Education

With the February 17, 2009 deadline for the digital transition approaching, and with the public launching of the converter box program in January 2008, Congressional concern is focusing on the adequacy of efforts to inform the public of the digital transition. A primary goal is preventing analog over-the-air households from losing television service in the event that these households do not purchase a converter box or take other measures to ensure the ability to receive digital broadcasts after February 17, 2009.

A survey conducted by the National Association of Broadcasters (NAB) found that 56% of over-the-air viewers have never seen, heard, or read anything about the digital transition, that only 10% were able to guess the right year when analog broadcasts will cease, and that only 1% to 3% knew that the transition would be complete by February 2009.[11]

A subsequent survey conducted by the Association of Public Television Stations (APTS) in August 2007 found that 51.3% of Americans were unaware of the DTV transition. A previous APTS survey in November 2006 found the percentage of Americans unaware of the DTV transition at 61.2%.[12]

Two federal agencies — the NTIA and the FCC — are directly engaged in consumer education efforts regarding the digital transition. Currently, the NTIA is statutorily funded (by P.L. 109-171, the Deficit Reduction Act of 2005) at "not more than $5,000,000 for consumer education concerning the digital television transition and the availability of the digital-to-analog converter box program." The NTIA's DTV consumer education efforts is focused on raising awareness of the coupon program, particularly with five target groups most likely to be affected by the digital transition: senior citizens, the economically disadvantaged, rural residents, people with disabilities, and minorities. To reach those groups and the American public in general, the NTIA is pursuing a strategy of leveraging its resources by partnering with private sector stakeholder groups representing those constituencies most at risk. NTIA is also working with the DTV Transition Coalition, a broad-based coalition of business, trade, and industry groups as well as grass roots and membership organizations. In addition to working with private sector groups, NTIA is working with federal government agencies that target economically disadvantaged Americans.[13]

Meanwhile, the Administration has requested $1.5 million for the FCC in FY2008 for DTV consumer education; the FY2008 House Financial Services and General Government Appropriations bill (H.R. 2829; H.Rept. 110-207), passed by the House on June 28, 2007, would provide $2 million. Similar to the NTIA, the FCC is pursuing collaborative partnerships with private and public sector entities to target outreach to vulnerable populations and to raise the general awareness of the American public about the DTV transition. The FCC has become a member of the DTV Transition Coalition, prepared and issued consumer publications and web materials, and is promoting DTV awareness by attending and holding events and conferences.[14]

The significant reliance of the FCC and the NTIA on the private sector for DTV public education has led some to question whether the federal government should assume a more proactive role in promoting DTV public education activities. On July 30, 2007, in response to criticisms and suggestions on DTV consumer education raised by a May 24, 2007 letter[15] from the House Energy and Commerce Committee, the FCC released a Notice of Proposed Rule Making (NPRM) on a DTV Consumer Education Initiative.[16] The NPRM requests public comments on a number of proposals to raise awareness among the public of the DTV transition, including broadcaster public service announcements, broadcaster consumer education reporting, multichannel video programming distributor (MVPD) customer bill notices, consumer electronics manufacturer notices, consumer electronics retailer reporting on its staff training, and other proposals.

Meanwhile, in testimony before the Senate Special Committee on Aging, the Government Accountability Office (GAO) stated that difficulties remain in implementing consumer education programs. GAO testified that because private sector DTV outreach efforts are voluntary, government cannot be assured of their extent and that "given the different interests represented by industry stakeholders, messages directed at consumers vary and might lead to confusion."[17] As requested by the House Committee on Energy and Commerce, GAO is performing an ongoing assessment of public and private sector DTV consumer education programs and is planning a series of consumer surveys leading up to the transition date.

A major component of any DTV public education campaign is likely to be the airing of public service announcements (PSAs). The National Association of Broadcasters is preparing PSAs to be delivered to local broadcasters by December 2007. It will be up to local broadcasters to decide when and how often to air the PSAs. Meanwhile, in September 2007, the National Cable and Telecommunications Association (NCTA) began running on cable channels a $200 million English and Spanish language advertising campaign on the digital transition; NCTA will continue the advertising spots through February 2009.[18] In its NPRM, the FCC states its belief that PSAs are the most effective and efficient way to reach over-the-air television viewers about the digital transition. The FCC is proposing to require television broadcast licensees to conduct on-air consumer education efforts and is asking for comments on the content of such PSAs, when and how often they should be run, whether similar requirements should be imposed on all broadcasters, and other related questions.[19]

KEY ISSUES

The Deficit Reduction Act of 2005 set a February 17, 2009 deadline for the digital transition and established a digital converter box coupon program to mitigate the switch-over costs to consumers with analog televisions. The key issue for Congress is the extent to which American households will be ready for the digital transition, and whether measures taken by the government and the private sector are sufficient to ensure that televisions across America do not "go dark" on February 18, 2009.

Two lines of inquiry have repeatedly been raised in Congressional hearings. First, are public education and outreach efforts sufficient, and is the federal government playing a sufficient role in leading that effort? With limited funding, both the FCC and the NTIA are relying heavily on a strategy of leveraging private sector efforts. On the one hand, private sector groups have a market incentive to ensure that the public is ready for the digital transition: for example, the consumer electronics industry wants to sell DTV products, and broadcasters want their viewers to be able to continue watching their local broadcasts. Accordingly, industry groups have begun to launch multifaceted public outreach campaigns. On the other hand, critics question whether market forces will ensure that public outreach efforts are sufficiently targeted to those segments of American society (the elderly, non-English speakers, rural areas, disabled citizens, minorities, the economically disadvantaged) that may be more at risk of being adversely affected by the digital switch-over. Critics also assert that industry outreach will likely reflect each industry sector's interests, and that a formal federal coordination and leadership effort is needed to ensure that a unified, consistent, and balanced message is conveyed to the public.

A second major question is the extent to which NTIA's converter box program will meet the needs of analog television households. According to the statute, households can begin requesting converter boxes as of January 1, 2008. Given that private sector participation in the converter box program is voluntary, will sufficient numbers of converter boxes be manufactured, and will retail outlets — whether large or small stores, whether in urban, suburban, or rural areas — stock sufficient numbers of boxes to meet the demand of consumers seeking to redeem the $40 coupons? Also, given that coupons will expire three months after households receive them, how effectively will NTIA be able to assess and monitor the balance of the demand for coupons with the local supply of converter boxes? How effectively will fraud, waste, and abuse be avoided and combated — particularly among vulnerable populations such as the elderly? And finally, will funding for the coupons — $990 million in the initial period and a possible additional $510 million in a contingent period — be sufficient to meet the total demand?

No definitive answers to these questions are possible until the converter box program is implemented and the digital transition proceeds. The best-case scenario is that public awareness of the digital transition will become ubiquitous during 2008, that converter boxes will be readily available to all who want them, and that the digital transition will proceed smoothly. The worst-case scenario is that public awareness of the digital transition will continue to lag and that converter boxes will not be uniformly available in retail outlets, leading to widespread confusion and frustration in many American households. At issue for Congress is whether the federal government's current programs and reliance on private sector stakeholders will lead to a successful digital transition with a minimum amount of disruption to American TV households or, alternatively, whether additional legislative measures are warranted.

ACTIVITIES IN THE 110TH CONGRESS

Congress is closely monitoring and overseeing federal and private sector efforts to ensure a digital transition that proceeds as smoothly as possible. The House Committee on Energy and Commerce, Subcommittee on Telecommunications and the Internet, held a hearing, "The Status of the Digital Television Transition," on March 28, 2007, the Senate Committee on Commerce, Science and Transportation held a hearing, "Preparing Consumers for the Digital Television Transition," on July 26, 2007, and the Senate Special Committee on Aging held a hearing, "Preparing for the Digital Television Transition: Will Seniors Be Left in the Dark?," on September 10, 2007. At these hearings, Members expressed concerns that federal and private sector efforts to educate the public about the DTV transition may be inadequate, that DTV education programs at the FCC and NTIA are underfunded, that converter boxes may not be uniformly available at retail outlets throughout the nation, and that the FCC is not exercising sufficient overall leadership of the DTV transition.

The following are DTV-related bills introduced into the 110th Congress:

H.R. 608 (Barton). Digital Television Consumer Education Act of 2007. Requires the FCC to create a DTV public education program, to convene a DTV Advisory Group to coordinate consumer outreach, and to report to Congress every six months on the progress of consumer education efforts. Requires NTIA to report to Congress every 90 days on the progress of the converter box coupon program. Requires retailers, cable and satellite operators, and broadcasters to take various measures to inform the public about the digital transition. Introduced January 22, 2007; referred to Committee on Energy and Commerce.

H.R. 2566 (Engel). National Digital Television Consumer Education Act. Requires TV retailers and distributors to place signs next to all analog TV displays with an advisory that a set-top box is necessary after February 17, 2009, to continue using the TV. Also requires broadcasters to air Public Service Announcements for more than a year before the transition to inform the public about the change and the set-top box subsidy program. Introduced June 5, 2007; referred to Committee on Energy and Commerce.

H.R. 2829 (Serrano). Financial Services and General Government Appropriations Act, 2008. House Appropriations Committee report H.Rept. 110-207, passed by the House on June 28, 2007, would provide $2 million to the FCC for DTV consumer education. Senate Appropriations Committee report (S.Rept. 110-129) does not address DTV. Placed on Senate Legislative Calendar, July 13, 2007.

H.R. 2917 (Butterfield). Transition Education Accountability Report Act of 2007. Requires the FCC to submit a report to Congress describing the measures taken by the FCC, NTIA, and other federal agencies to inform the public of the transition to digital television. Introduced June 28, 2007; referred to Committee on Energy and Commerce.

S. 2125 (Kohl). Preparing America's Seniors for the Digital Television Transition Act of 2007. Establishes an interagency federal taskforce to educate older Americans on the DTV transition. Requires retailers, cable and satellite operators, and broadcasters to take various measures to inform the public about the digital transition. Directs the FCC to award grants for DTV public education. Requires modifications in the digital-to-analog converter box program. Requires the NTIA and the FCC to provide 90-day progress reports to Congress. Introduced October 2, 2007; referred to Committee on Commerce, Science and Transportation.

FOR FURTHER INFORMATION

A variety of websites have been established to provide basic information to consumers on many aspects of the digital transition. The following is a partial listing.

Federal Communications Commission (FCC) [http://www.dtv.gov]

National Telecommunications and Information Administration (NTIA) [http://www.ntia. doc.gov/dtvcoupon/index.html]

DTV Transition Coalition [http://www.dtvtransition.org/]

National Association of Broadcasters (NAB) [http://www.dtvanswers.com/]

Consumer Electronics Retailers Coalition (CERC) [http://www.ceretailers.org/transtodtv.htm]

Consumer Electronics Association (CEA) [http://www.myceknowhow.com/ digital Television.cfm]

National Cable and Telecommunications Association (NCTA) [http://www. getreadyfordigitaltv.com/]

Satellite Broadcasting and Communications Association of America (SBCA) [http://www.sbca.com/hdtv_index.asp]

REFERENCES

[1] As of March 1, 2007, *all* analog televisions manufactured, imported, or shipped across state lines are required to have a built-in digital tuner, and will therefore not require a converter box. Retailers are permitted to sell analog-only devices from existing inventory, but are required by the FCC to display a "consumer alert" label explaining that the device will require a converter box in order to receive over-the-air television signals after February 17, 2009.

[2] The February 17, 2009 deadline applies only to full power television stations. Low power television stations, including Class A stations and translator stations, will transition to digital broadcasting at a date yet to be determined by the FCC.

[3] Testimony of K. James Yager on behalf of the National Association of Broadcasters and the Association for Maximum Service Television, hearing before the House Committee on Energy and Commerce, Subcommittee on Telecommunications and the Internet, March 28, 2007, p. 11. Available at [http://energycommerce.house.gov/ cmte_mtgs/ 110-ti-hrg.032807.Yager-testimony.pdf]

[4] Letter from Jack Sander, Joint Board Chair, National Association of Broadcasters, to the Honorable Kevin Martin, Chairman, FCC, Re: *In the Matter of DTV Consumer Education Initiative*, MB Docket No. 07-148, p. 1.

[5] *National Journal's Technology Daily*, PM Edition, March 16, 2007, Vol. 10, No. 9

[6] Association of Public Television Stations, "APTS Study Shows Older Americans Less Prepared for the Digital TV Transition," Press Release, July 24, 2007.

[7] U.S. Government Accountability Office, Testimony before the Subcommittee on Telecommunications and the Internet, Committee on Energy and Commerce, House of Representatives, *Digital Broadcast Television Transition: Estimated Cost of Supporting Set-Top Boxes to Help Advance the DTV Transition*, February 17, 2005, p. 7-8.

[8] U.S. Department of Commerce, National Telecommunications and Information Administration, ""Rules to Implement and Administer a Coupon Program for Digital-to-Analog Converter Boxes," 47 CFR 301, *Federal Register*, Vol. 72, No. 51, March 15, 2007, pp. 12097-12121.

[9] National Telecommunications and Information Administration, *DTV Converter Box Program Information Sheet for Manufacturers*, March 2007, available at [http://www.ntia.doc.gov/dtvcoupon/DTVmanufacturers.pdf].

[10] National Telecommunications and Information Administration, *DTV Converter Box Program Information Sheet for Retailers*, September 2007, available at [http://www.ntia.doc.gov/dtvcoupon/DTVretailers.pdf].

[11] Testimony of K. James Yager on behalf of the National Association of Broadcasters and the Association for Maximum Service Television, hearing before the House Committee on Energy and Commerce, Subcommittee on Telecommunications and the Internet, March 28, 2007, p. 14. Available at [http://energycommerce.house. gov/cmte_mtgs/ 110-ti-hrg.032807.Yager-testimony.pdf].

[12] Association of Public Television Stations, Press Release, "Government Gets Failing Grade on DTV Transition," September 24, 2007. Available at [http://www.apts.org/ news/govfailinggrade.cfm].

[13] For information on NTIA DTV consumer education efforts, see Testimony of John Kneuer, Assistant Secretary for Communications and Information, National Telecommunications and Information Administration, hearings held by the Senate Committee on Commerce, Science and Transportation, "Preparing Consumers for the Digital Television Transition," July 26, 2007. Available at [http://commerce.senate.gov/ public/_files/JohnMRKneuerTestimonyv2.pdf].

[14] Testimony of Catherine Seidel, Chief, Consumer and Governmental Affairs Bureau, Federal Communications Commission, hearings held by the Senate Committee on Commerce, Science and Transportation, "Preparing Consumers for the Digital Television Transition," July 26, 2007. Available at [http://commerce.senate.gov/ public/_files/ WrittenStatementofCathySeidel7262007Hearing.pdf].

[15] Available at [energycommerce.house.gov/Press_110/FCC.052407.Martin.ltr.DTV.pdf].

[16] FCC, Notice of Proposed Rulemaking, In the Matter of DTV Consumer Education Initiative, MB Docket No. 07-148, FCC 07-128, 22 p.

[17] Government Accountability Office, Testimony Before the Senate Special Committee on Aging, *Digital Television Transition: Preliminary Information on Initial Consumer Education Efforts*, GAO-07-1248T, September 19, 2007, p. 9. Available at [http://www.gao.gov/new.items/d071248t.pdf]

[18] National Cable and Telecommunications Association, Press Release, "Cable Launches $200 Million Digital TV Transition Consumer Education Campaign," September 6, 2007.

[19] Ibid., p. 3.

In: Telecommunications and Media Issues
Editors: A. N. Moller, C. E. Pletson, pp. 75-86

ISBN: 978-1-60456-294-1
© 2008 Nova Science Publishers, Inc.

Chapter 5

THE V-CHIP AND TV RATINGS: MONITORING CHILDREN'S ACCESS TO TV PROGRAMMING[*]

Patricia Moloney Figliola

ABSTRACT

To assist parents in supervising the television viewing habits of their children, the Communications Act of 1934 (as amended by the Telecommunications Act of 1996) requires that, as of January 1, 2000, new television sets with screens 13 inches or larger sold in the United States be equipped with a "V-chip" to control access to programming that parents find objectionable. Use of the V-chip is optional. In March 1998, the Federal Communications Commission (FCC) adopted the industry-developed ratings system to be used in conjunction with the V-chip. Congress and the FCC have continued monitoring implementation of the V-chip. Some are concerned that it is not effective in curbing the amount of TV violence viewed by children and want further legislation.

In July 2004, the FCC initiated a Notice of Inquiry (NOI) to seek comments relating to the "presentation of violent programing and its impact on children." The Report in this proceeding was released by the FCC on April 25, 2007. In the report, the FCC, among other findings, (1) found that on balance, research provides strong evidence that exposure to violence in the media can increase aggressive behavior in children, at least in the short term; (2) stated that the V-chip is of limited effectiveness in protecting children from violent television content and observed that cable operator-provided advanced parental controls do not appear to be available on a sufficient number of cable-connected television sets to be considered an effective solution at this time; and (3) found that studies and surveys demonstrate that the voluntary TV ratings system is of limited effectiveness in protecting children from violent television content.

Congress may wish to consider a number of possible options to support parents in controlling their children's access to certain programming. Some of these options would require only further educational outreach to parents, while others would require at least regulatory, if not legislative, action. Specifically, Congress may wish to consider ways to promote awareness of the V-chip and the ratings system; whether the current set of

---

[*] Excerpted from CRS Report RL32729, dated July 17, 2007.

media-specific ratings will remain viable in the future or whether a uniform system would better serve the needs of consumers; and whether independent ratings systems and an "open" V-chip that would allow consumers to select the ratings systems they use would be more appropriate than the current system.

BACKGROUND

Recent research indicates that 89% of parents are "somewhat" to "very" concerned that "their children are being exposed to too much inappropriate content in entertainment media."[1] Further, parents cited television as the medium that caused them the most concern.[2] Although exposure to inappropriate material has long been a concern to parents, only since the Telecommunications Act of 1996[3] has there been a nationwide effort to provide parents with a tool to control their children's television viewing — the V-chip.[4]

The V-chip, which reads an electronic code transmitted with the television signal (cable or broadcast),[5] is used in conjunction with a television programming rating system. Using a remote control, parents can enter a password and then program into the television set which ratings are acceptable and which are unacceptable. The chip automatically blocks the display of any programs deemed unacceptable; use of the V-chip by parents is entirely optional.[6]

As of January 1, 2000, all new television sets with a picture screen 13 inches or greater sold in the United States must be equipped with the V-chip.[7] Additionally, some companies also offer devices that can work with non-V-chip TV sets.

DEVELOPMENT OF THE V-CHIP RATINGS SYSTEM

The initial ratings system was developed during 1996 and 1997, but encountered criticism from within Congress as well as from groups such as the National Parent-Teacher Association. In response to those concerns, an expanded ratings system was adopted on July 10, 1997, and went into effect October 1, 1997.

Initial Ratings System

The first step in implementing the mandate of the law was to create a ratings system for television programs, analogous to the one developed and adopted for movies by the Motion Picture Association of America (MPAA) in 1968. The law urged the television industry to develop a voluntary ratings system acceptable to the FCC, and the rules for transmitting the rating, within one year of enactment. The ratings system is intended to convey information regarding sexual, violent or other indecent material about which parents should be informed before it is displayed to children, provided that nothing in the law should be construed to authorize any rating of video programming on the basis of its political or religious content.

After initial opposition, media and entertainment industry executives met with then-President Clinton on February 29, 1996, and agreed to develop the ratings system because of political pressure to do so. Many in the television industry were opposed to the V-chip, fearing that it would reduce viewership and reduce advertising revenues. They also questioned whether it violated the First Amendment. Industry executives said they would not challenge the law immediately, but left the option open should they deem it necessary.

Beginning in March 1996, a group of television industry executives[8] under the leadership of Jack Valenti, then-President of the MPAA (and a leader in creating the movie ratings), met to develop a TV ratings system. On December 19, 1996, the group proposed six age-based ratings (TV-Y, TV-Y7, TV-G, TV-PG, TV-14 and TV-M), including text explanations of what each represented in terms of program content. In January 1997, the ratings began appearing in the upper left-hand corner of TV screens for 15 seconds at the beginning of programs, and were published in some television guides. Thus, the ratings system was used even before V-chips were installed in new TV sets.

Ratings are assigned to shows by the TV Parental Guidelines Monitoring Board. The board has a chairman and six members each from the broadcast television industry, the cable industry, and the program production community. The chairman also selects five non-industry members from the advocacy community, for a total of 24 members.

News shows and sports programming are not rated. Local broadcast affiliates may override the rating given a particular show and assign it another rating.

The Current "S-V-L-D" Ratings System

Critics of the initial ratings system argued that the ratings provided no information on why a particular program received a certain rating. Some advocated an "S-V-L" system (sex, violence, language) to indicate with letters why a program received a particular rating, possibly with a numeric indicator or jointly with an age-based rating. Another alternative was the Home Box Office/Showtime system of ten ratings such as MV (mild violence), V (violence), and GV (graphic violence).

In response to the criticism, most of the television industry agreed to a revised ratings system (see box, below) on July 10, 1997, that went into effect October 1, 1997. The revised ratings system added designators to indicate whether a program received a particular rating because of sex (S), violence (V), language (L), or suggestive dialogue (D). A designator for fantasy violence (FV) was added for children's programming in the TV-Y7 category. On March 12, 1998, the FCC approved the revised ratings system, along with V-chip technical standards, and the effective date for installing them.[9]

In May 1999, the FCC created a V-chip Task Force, chaired by then-Commissioner Gloria Tristani. Among other things, the task force was charged with ensuring that the blocking technology was available and that ratings were being transmitted ("encoded") with TV programs; educating parents about V-chip; and gathering information on the availability, usage, and effectiveness of the V-chip. The task force issued several reports and surveys.[10] A February 2000 task force survey found that most broadcast, cable, and premium cable networks, and syndicators, were transmitting ratings ("encoding") and those that were not either planned to do so in the near future or were exempt sports or news networks. Of the major broadcast and cable networks, only NBC and Black Entertainment Television do not use the S-V-L-D indicators, using the original ratings system instead.

Table 1. U.S. Television Industry's Revised TV Ratings System

TV Y	**TV-Y All Children** This program is designed to be appropriate for all children. Whether animated or live-action, the themes and elements in this program are specifically designed for a very young audience, including children from ages 2-6. This program is not expected to frighten younger children.
TV Y7	**TV-Y7 Directed to Older Children** This program is designed for children age 7 and above. It may be more appropriate for children who have acquired the developmental skills needed to distinguish between make-believe and reality. Themes and elements in this program may include mild fantasy or comedic violence, or may frighten children under the age of 7. Therefore, parents may wish to consider the suitability of this program for their very young children.
TV Y7 FV	**TV-Y7-FV Directed to Older Children-Fantasy Violence** For those programs where fantasy violence may be more intense or more combative than other programs in the TV-Y7 category, such programs will be designated TV-Y7-FV.
TV G	**TV-G General Audience** Most parents would find this program appropriate for all ages. Although this rating does not signify a program designed specifically for children, most parents may let younger children watch this program unattended. It contains little or no violence, no strong language and little or no sexual dialogue or situations.
TV PG	**TV-PG Parental Guidance Suggested** This program contains material that parents may find unsuitable for younger children. Many parents may want to watch it with their younger children. The theme itself may call for parental guidance and/or the program contains one or more of the following: moderate violence (V), some sexual situations (S), infrequent coarse language (L), or some suggestive dialogue (D).
TV 14	**TV-14 Parents Strongly Cautioned** This program contains some material that many parents would find unsuitable for children under 14 years of age. Parents are strongly urged to exercise greater care in monitoring this program and are cautioned against letting children under the age of 14 watch unattended. This program contains one or more of the following: intense violence (V), intense sexual situations (S), strong coarse language (L), or intensely suggestive dialogue (D).
TV MA	**TV-MA Mature Audience Only** This program is specifically designed to be viewed by adults and therefore may be unsuitable for children under 17. This program contains one or more of the following: graphic violence (V), explicit sexual activity (S), or crude indecent language (L).

FEDERAL COMMUNICATIONS COMMISSION ACTION

On April 25, 2007, the FCC released a report on the "presentation of violent programing and its impact on children."[11] In the report, the FCC —

- found that on balance, research provides strong evidence that exposure to violence in the media can increase aggressive behavior in children, at least in the short term;
- noted that although viewer-initiated blocking and mandatory ratings would impose lesser burdens on protected speech, skepticism remains that they will fully serve the government's interests in promoting parental supervision and protecting the well-being of minors;
- stated that the V-chip is of limited effectiveness in protecting children from violent television content;
- observed that cable operator-provided advanced parental controls do not appear to be available on a sufficient number of cable-connected television sets to be considered an effective solution at this time;
- stated that further action to enable viewer-initiated blocking of violent television content would serve the government's interests in protecting the well-being of children and facilitating parental supervision and would be reasonably likely to be upheld as constitutional;
- found that studies and surveys demonstrate that the voluntary TV ratings system is of limited effectiveness in protecting children from violent television content;
- stated that Congress could develop an appropriate definition of excessively violent programming, but such language needs to be narrowly tailored and in conformance with judicial precedent;
- suggested that industry could on its own initiative commit itself to reducing the amount of excessively violent programming viewed by children (e.g., broadcasters could adopt a family hour at the beginning of prime time, during which they decline to air violent content);
- observed that multichannel video programming providers (MVPDs) could provide consumers greater choice in how they purchase their programming so that they could avoid violent programming. (e.g., an a la carte regime would enable viewers to buy their television channels individually or in smaller bundles); and
- found that Congress could implement a time channeling solution and/or mandate some other form of consumer choice in obtaining video programming, such as the provision by MVPDs of video channels provided on family tiers or on an a la carte basis (e.g., channel blocking and reimbursement).

CONGRESSIONAL ACTION

Since 2003, the television industry and the FCC faced increasing scrutiny for what was perceived by many in Congress, as well as the public, as a sharp increase in the amount of indecent programming. Two of the most notable events that have taken place with respect to this issue were the FCC's determination that the use of the "f-word" by an artist during an award ceremony was not indecent and, four days later, an incident during the Super Bowl XXXVIII half-time show that included a performance in which one of the entertainer's breasts was revealed.

110th Congress

Thus far, there has been no legislation introduced specifically on the issue of media violence or the V-chip; however, the Senate and the House of Representatives have each held one hearing:

- The Senate Committee on Commerce, Science, and Transportation held a hearing, "Impact of Media Violence on Children," on June 26, 2007.[12] The hearing focused on issues related to the impact of violent television programming on children, including issues raised by the FCC report, "Violent Television Programming And Its Impact On Children."
- The House Committee on Energy and Commerce Subcommittee on Telecommunications and the Internet held a hearing, "Images Kids See on the Screen," on June 22, 2007.[13] The hearing included discussion of advertising for junk food aimed at children and on the inability of the V-chip to screen out undesirable advertising.

Also, media violence and the V-chip were both discussed during an April 17, 2007 House Committee on Appropriations, Subcommittee on Financial Services and General Government hearing on the FCC's FY2008 budget request.

109th Congress

During the 109th Congress, Senators John Rockefeller and Kay Bailey Hutchison introduced S. 616, the Indecent and Gratuitous and Excessively Violent Programming Control Act, on March 14, 2005. Specifically with respect to the television ratings and the V-chip, S. 616 would have required the FCC to assess the effectiveness of both the ratings system and the V-chip and report annually on it's findings. Further, if the FCC were to find that the ratings system and V-chip did not adequately protect children from excessive violent and sexual content, it would be required to undertake a rulemaking to require broadcasters to do more to protect children from such content, including whether to use a new system not developed by the industry. The bill would also have required more consistent and meaningful labeling of violent and sexual content, to include re-broadcasting such labeling for 30 seconds every 30 seconds, whether the programming was received via broadcast, cable, or satellite. S. 616 was referred to the Committee on Commerce, Science, and Transportation on March 14, 2005; no further action was taken.[14]

EFFECTIVENESS OF THE V-CHIP: CURRENT RESEARCH

Since 1998, the Kaiser Family Foundation (KFF) has conducted ongoing research into the impact of media violence on children and the effectiveness of the V-chip and television ratings as tools for parents to control access to undesirable television content.[15] In the Foundation's most recent report, "Parents, Media, and Public Policy: A Kaiser Family

Foundation Survey," (KFF Study)[16] a majority of parents reported that they were "very" concerned about the amount of sex (60%) and violence (53%) their children are exposed to on TV.[17]

Overall, the parents interviewed for the study stated that they were more concerned about inappropriate content on TV than in other media: 34% said TV concerned them most, compared to 16% who said the Internet, 10% movies, 7% music, and 5% video games. Half (50%) of all parents said they have used the TV ratings to help guide their children's viewing, including one in four (24%) who said they use them "often."[18]

Furthermore, the study revealed that while use of the V-chip has increased substantially since 2001, when 7% of all parents said they used it, it remains modest at just 15% of all parents, or about four in 10 (42%) of those who have a V-chip in their television and know it. Nearly two-thirds (61%) of parents who have used the V-chip said they found it "very" useful.[19]

Other significant findings reported included:

- After being read arguments on both sides of the issue, nearly two-thirds of parents (63%) said they favored new regulations to limit the amount of sex and violence in TV shows during the early evening hours, when children were most likely to be watching (35% are opposed).[20]
- A majority (55%) of parents said ratings should be displayed more prominently and 57% said they would rather keep the current rating systems than switch to a single rating for TV, movies, video games, and music (34% favor the single rating).[21]
- When read the competing arguments for subjecting cable TV to the same content standards as broadcasters, half of all parents (52%) said that cable should be treated the same, while 43% said it should not.[22]
- Most parents who have used the TV ratings said they found them either "very" (38%) or "somewhat" (50%) useful.[23]
- About half (52%) of all parents said most TV shows are rated accurately, while about four in ten (39%) said most are not.[24]
- Many parents do not understand what the various ratings guidelines mean. For example, 28% of parents of young children (2-6 years old) knew what the rating TV-Y7 meant (directed to children age 7 and older) while 13% thought it meant the opposite (directed to children under 7); and only 12% knew that the rating FV ("fantasy violence") is related to violent content, while 8% thought it meant "family viewing."[25]

In releasing the survey results, Vicky Rideout, Vice President and Director of the Kaiser Family Foundation's Program for the Study of Entertainment Media and Health, commented, "While many parents have used the ratings or the V-chip, too many still don't know what the ratings mean or even that their TV includes a V-chip."[26]

A number of groups conducted research and published opinion pieces questioning the usefulness and/or legality of the V-chip and the ratings system after the 1996 Telecommunications Act was enacted (e.g., the American Civil Liberties Union, Cato Institute, Morality in Media). Since that time, opposition has waned and even the recent controversies did not renew it. Further, while the V-chip and the ratings system can block objectionable or indecent programming when used in tandem, since the incidents were broadcast "live" and did not have ratings that would have blocked them, neither the V-chip nor the ratings system would have been effective in either case. Therefore, some could claim that the V-chip and the ratings system, while useful tools in many cases, remain unreliable tools for parents because they cannot guarantee all objectionable content will be blocked.

ISSUES FOR CONGRESS

Congress may wish to consider a number of possible options to support parents in monitoring and controlling their children's access to certain programming. Some of these options would require only further educational outreach to parents, while others would require at least regulatory, if not legislative action.

Awareness of the V-Chip and the Ratings System

According the 2004 KFF Study, parents also indicated that they would like to see the ratings displayed more prominently to make it easier to notice them. Such findings are consistent with a lack of wide-spread usage or even awareness of the V-chip. Specifically, as noted above, the 2004 KFF study indicated that even after years of being available, only 42% of parents who have a V-chip and are aware of it actually use it. However, of the parents that had used the V-chip, 89% found it "somewhat" to "very" useful.[27] Those figures would indicate that increased knowledge of the V-chip would substantially increase parents' perceptions of control over their children's television viewing.

One of the easiest approaches to increasing the use of the V-chip may likely be to step up parental awareness programs through, for example, public service announcements on television, educational materials on the FCC website, and possibly pubic service advertisements in print media. Additionally, such educational materials could be made available on Congressional Member websites for constituents to download. Such actions would not require any new legislation or additional work by the ratings board or related entities; however, some initially may require funding.

Media-Specific vs. Uniform Ratings

One of the ongoing issues related to the use of the V-chip is that, according to the KFF study, only about half of parents actually use the television ratings. That is low in comparison with the movie ratings, which are used by approximately 78% of parents, but in line with the use of ratings for music and video games.[28] One contributor to the low use of the ratings is likely that so few parents actually understand the ratings. For example, as stated earlier, only 12% of parents of young children knew that "FV" is the rating for Fantasy Violence; further, 8% believed it to mean "Family Viewing." As noted by the researchers in their report, the FV rating "is the only rating that denotes anything about the violent content of children's programming, one of the impetuses for the development of the ratings system" in the first place. Finally, overall, 20% of parents had never even heard of the ratings system.[29]

In light of those figures, it could appear that parents might prefer a single, unified ratings system that would be applied across different media. However, while 34% of parents said they would prefer a unified system, 57% opposed a unified system.[30] Given the overall findings by KFF regarding parents' knowledge and use of the ratings system, there appears to be enough ambiguity on this issue to warrant further investigation by Congress.

Independent Ratings Systems and an "Open" V-Chip

Under current legislative and regulatory mandates, the V-chip is only required to "read" the TV Parental Guidelines and the MPAA (movie) Ratings. This means that any independent system can only be used to augment parental knowledge, not to program the V-chip. So, while a range of varied, independent ratings systems can serve to provide additional information to parents, they cannot be used with the current closed V-chip technology. In order for these ratings to become as useful as possible, the V-chip would have to be able to read them.

The opportunity to encourage the further development of private ratings systems exists in the transition to digital television. Beginning in April 2005, all broadcasters must simulcast 100 percent of their National Television System Committee (commonly referred to as "NTSC") programming on their digital channel; by the end of 2006, broadcasters must turn off their analog signal.[31] Through either regulatory (i.e., FCC) or legislative action, television manufacturers could be required to install an open V-chip that could be reprogrammed to read altered or even completely new ratings. An "open" V-chip requirement would allow changes to the current system to be read as well as accommodate any other ratings system(s). This issue is currently under consideration at the FCC.[32]

RELATED READING

Other Reports and Documents

"Parents, Media, and Public Policy: A Kaiser Family Foundation Survey," Kaiser Family Foundation, Fall 2004, [http://www.kff.org/entmedia/ entmedia092304pkg.cfm].
"V-chip Frequently Asked Questions," Children Now, [http://www.childrennow.org/media/ vchip/vchip-faq.html].
"Summary of Focus Group Research on Media Ratings Systems," A Study Commissioned by PSV Ratings, Inc., Spring 2003, [http://www.independentratings.org/Parents_Views.pdf].

Websites

Federal Communications Commission V-chip Information, [http://www.fcc.gov/vchip/].

REFERENCES

[1] "Parents, Media, and Public Policy: A Kaiser Family Foundation Survey," Kaiser Family Foundation, Fall 2004, p. 2. Available online at [http://www.kff.org/entmedia/ entmedia092304pkg.cfm]. (KFF Study)

[2] KFF Study, p. 2. Specifically, 63% said they were ""very concerned" and 26% said they were "somewhat concerned."

[3] Telecommunications Act of 1996, P.L. 104-104, February 8, 1996, available online at [http://www.fcc.gov/Reports/ 1934new.pdf]. The 1996 Act amended the Communications Act of 1934 (47 U.S.C. 101, *et seq.*), updating some existing sections and adding new sections to account for new technologies. One such addition to the law was to mandate the inclusion of a computer chip in new television sets to allow parents more control over the programming viewed by their children (47 U.S.C. 303 (x)). The 1934 Act, as amended by the 1996 Act, is available online at [http://www.fcc.gov/Reports/tcom1996.pdf].

[4] Although commonly believed to be short for "violence," the V in V-chip is actually short for "ViewControl," the name given by the inventor of the device. See "V-Chip Technology Invented by Professor Tim Collings," available online at [http://www.tri-vision.ca/ documents/Collings%20As%20Inventor.pdf]. See also, "The History of Invention," available online at [http://www.cbc.ca/kids/general/the-lab/ history-of-invention/vchip.html].

[5] The ratings data are sent on line 21 of the Vertical Blanking Interval found in the National Television System Committee (NTSC) signals used for U.S. television broadcasting.

[6] This report focuses on the use of the V-chip and the ratings system as a tool to assist parents in selecting appropriate television programming for their children. However, both the V-chip and the ratings system can be used by a wide range of viewers, from individuals who, themselves, do not wish to view content they find objectionable to individuals who may be babysitting on an intermittent basis in their homes. Further, the V-chip and the television ratings are closely related to another issue — that of broadcast indecency and how to define and enforce the appropriate use of the public airwaves by the television media. That issue is discussed in greater detail in CRS Report RL32222, *Regulation of Broadcast Indecency: Background and Legal Analysis*, by Henry Cohen.

[7] 47 U.S.C. 303(x).

[8] The group included the national broadcast networks; independent, affiliated and public television stations; cable programmers; producers and distributors of cable programming; entertainment and movie studios; and members of the guilds representing writers, directors, producers and actors.

[9] As of January 1, 2000, all new television sets with a picture screen 13 inches or greater sold in the United States must be equipped with the V-chip.

[10] See [http://www.fcc.gov/vchip].

[11] *In the Matter of Violent Television Programming and its Impact on Children* (FCC 04-175, MB Docket 04-261), Notice of Inquiry (NOI), Adopted July 15, 2004; Released July 28, 2004. The NOI is available online at [http://hraunfoss.fcc.gov/edocs_public/ attachmatch/FCC-07-50A1.pdf].

[12] The hearing webpage containing witness statements and the archived video is online at [http://commerce.senate.gov/public/index.cfm?FuseAction=Hearings. Hearing and Hearing_ID=1879]. See also, Broadcasting and Cable, "TV Hammered In Violence Hearing," by John Eggerton, June 26, 2007. Available online at [http://www.broadcastingcable.com/article/CA6455579.html].

[13] The hearing webpage containing witness statements and the archived video is online at [http://energycommerce.house.gov/cmte_mtgs/110-ti-hrg.062207.ImagesKidsSEE.shtml]. See also, Broadcasting and Cable, "Markey Calls V-chip Limited Success," by John Eggerton, June 22, 2007. Available online at [http://www.broadcastingcable.com/article/ CA6454591.html].

[14] This bill also contains a measure related to increasing fines for violating rules on indecent programming, but that issue is outside the scope of this report. For more information on that topic, please refer to CRS Report RL32222, *Regulation of Broadcast Indecency: Background and Legal Analysis*, by Henry Cohen.

[15] See Kaiser Family Foundation, Program on Study of Entertainment Media and Health: Television/Video, [http://www.kff.org/entmedia/tv.cfm].

[16] "Parents, Media, and Public Policy: A Kaiser Family Foundation Survey," Kaiser Family Foundation, Fall 2004 (KFF Study). The survey of 1,001 parents of children ages 2-17 was conducted in July and August 2004.

[17] KFF Study, p. 3.

[18] KFF Study, p. 2.

[19] KFF Study, p. 7.

[20] KFF Study, p. 8.

[21] Ibid.

[22] Ibid.

[23] KFF Study, p. 5.

[24] Ibid.

[25] KFF Study, p. 6.

[26] KFF News Release, "Parents Favor New Limits on TV Content in Early Evening Hours; Half of Parents Say Cable TV Should Adhere to Same Standards as Broadcast TV; Use of the V-Chip is Up," September 23, 2004. Available online at [http://www.kff.org/entmedia/ entmedia092304nr.cfm].

[27] KFF Study, p. 7.

[28] KFF Study, p. 4.

[29] KFF Study, p. 6.

[30] KFF Study, p. 8.

[31] The December 31, 2006, deadline may be extended under a number of circumstances, detailed in CRS Report RL31260, Digital Television: An Overview, by Lennard Kruger.

[32] *In the Matter of Second Periodic Review of the Commission's Rules and Policies Affecting the Conversion To Digital Television*, MB Docket No. 03-15, RM 9832, Report and Order, September 4, 2004, paras. 154-159. Available online at [http://hraunfoss.fcc.gov/ edocs_public/attachmatch/FCC-04-192A1.pdf]. One issue that remains under consideration involves new language concerning the V-chip and how it will be incorporated into digital television sets. The Consumer Electronics Association (CEA) filed a petition to change the language that the FCC adopted in the

Order. That petition is available online at [http://hraunfoss.fcc.gov/edocs_public/ attachmatch/DA-03-1292A1.pdf]. The opposition to CEA's Petition for Reconsideration by Tri-Vision (the inventor of the V-chip) is available online at [http://www.tri-vision.ca/documents/2004/FCC%20Tri-vision%20Opposition.pdf]. See also *In the Matter of Children's Television Obligations of Digital Television Broadcasters*, MM Docket No. 00-167, Report and Order and Further Notice of Proposed Rulemaking, November 23, 2004, paras. 62-65. Available online at [http://hraunfoss.fcc.gov/ edocs_public/attachmatch/FCC-04-221A1.pdf].

In: Telecommunications and Media Issues
Editors: A. N. Moller, C. E. Pletson, pp. 87-95

ISBN: 978-1-60456-294-1
© 2008 Nova Science Publishers, Inc.

Chapter 6

COPYRIGHT PROTECTION OF DIGITAL TELEVISION: THE BROADCAST VIDEO FLAG[*]

Brian T. Yeh

ABSTRACT

In November 2003, the Federal Communications Commission (FCC) adopted a rule that required all digital devices capable of receiving digital television (DTV) broadcasts over the air, and sold after July 1, 2005, to incorporate technology that would recognize and abide by the broadcast video flag, a content-protection signal that broadcasters may choose to embed into a digital broadcast transmission as a way to prevent unauthorized redistribution of DTV content. However, in October 2004, the American Library Association and eight organizations representing a large number of libraries and consumers filed a lawsuit that challenged the power of the FCC to promulgate such a rule. In May 2005, the United States Court of Appeals for the District of Columbia Circuit ruled in *American Library Association v. Federal Communications Commission* that the FCC had exceeded the scope of its delegated authority in imposing the broadcast flag regime, and the court thus reversed and vacated the FCC's broadcast flag order.

Parties holding a copyright interest in content transmitted through DTV broadcasts, in particular broadcasters and television program creators, remain concerned about the unauthorized distribution and reproduction of copyrighted DTV content and thus continue to advocate the adoption of a broadcast video flag. However, several consumer, educational, and technology groups raise objections to the broadcast flag because, in their view, it would place technological, financial, and regulatory burdens that may stifle innovation, limit the consumer's ability to use DTV broadcasts in accordance with the Copyright Act's "fair use" principles, and possibly frustrate the use of digital television content by educators and librarians in distance education programs.

This report provides a brief explanation of the broadcast video flag and its relationship to digital television and summarizes the *American Library Association* judicial opinion. The report also examines a legislative proposal introduced in the 109th Congress, the Digital Content Protection Act of 2006, which appeared as portions of two bills, S. 2686 and H.R. 5252 (as reported in the Senate), that would have expressly granted statutory authority to the FCC under the Communications Act of 1934 to

[*] Excerpted from CRS Report RL33797, dated January 11, 2007.

promulgate regulations implementing a broadcast video flag system. Although not enacted, these bills represent approaches to authorizing the broadcast video flag system that may be of interest to the 110th Congress.

INTRODUCTION

Technological advances, a looming statutory deadline, and the need to reclaim analog spectrum occupied by television broadcasters have put digital television (DTV) on a fast track. At the same time, development of digital television has necessitated balancing the competing interests of content holders and consumer and technological industries. Reconciling these interests has led to the development of a broadcast video flag to combat unauthorized redistribution of content broadcast through digital television signals.[1] The move to protect digital content has been given urgency by the Federal Communications Commission's (FCC's) determination that broadcast transmissions be digital by December 31, 2006.[2] The 105th Congress, in the Balanced Budget Act of 1997, P.L. 105-33, made this date statutory.[3] However, the lack of widespread purchase and adoption by consumers of digital television equipment prompted the 109th Congress to extend the 2006 deadline; a provision of the Deficit Reduction Act of 2005, P.L. 109-171, established a "firm deadline" of February 17, 2009, for the digital transition.

What Is DTV?

Digital Television is a new television service representing the most significant development in television technology since the advent of color television in the 1950s. Three major components of DTV service must be present for consumers to enjoy a fully realized high-definition television viewing experience. First, digital programming must be available. Digital programming is content assembled with digital cameras and other digital production equipment. Second, digital programming must be delivered to the consumer via a digital signal. Third, consumers must have digital television equipment capable of receiving the digital signal and displaying digital programming for viewing.[4]

Developing a protocol for transmitting and receiving digital television in a way that accommodated competing interests has proved challenging. Digital content, like other media, can be relatively easily duplicated and distributed, especially with the aid of the Internet.[5] Unlike other types of content, duplication of digital information does not degrade the original. Whereas the quality of a VHS tape degrades after successive copies, a DVD may be copied almost infinitely with no effect on the quality of the medium. It is because of the ease and inexhaustible potential of copying digital media, coupled with the proliferation of Internet peer-to-peer file-sharing services, that content providers have greeted this new technology with some trepidation.

The Broadcast Video Flag

The broadcast video flag is a combination of technical specifications and federal regulations designed to combat unauthorized redistribution of content broadcast through digital television signals. Its adoption was prompted largely by the FCC's determination that broadcast transmissions be digital by December 31, 2006[6] (a deadline that has since been extended by Congress to February 17, 2009). The FCC imposed a transition to DTV in part to capitalize on the sharper images, CD-quality sound, and wider screen angles that are available from advanced digital technologies. However, in addition to the technological impetus, the FCC also has been motivated by the knowledge that broadcasters, upon receiving digital spectrum allotments, must relinquish their analog spectrum allotments to the FCC. The analog spectrum will in turn be auctioned for other commercial and public interests. Content providers, fearing widespread piracy that would endanger aftermarket sales (such as cable rebroadcast and DVD sales), urged the FCC to provide for a means to protect their assets. Meanwhile, consumer electronics and information technologists, as well as consumer rights groups, came together in an effort to minimize the possible negative outcome that a wide-scale regulation might have imposed.

The technical specifications behind the broadcast video flag were a compromise measure, premised on an understanding that more restrictive approaches (such as encrypted signals created at the source of transmission) imposed economically or technologically infeasible conditions. The compromise came after a consortium of content providers and consumer electronics and information technology groups came together, forming the Broadcast Protection Discussion Group (BPDG).[7] The result of this consortium was a *Final Report* published in June 2002, which was delivered to Representative Billy Tauzin, then-Chairman of the House Committee on Energy and Commerce. The report suggested a set of "robustness and compliance" rules for devices capable of demodulating digital television signals, which would require that such devices protect "flagged" content from being recorded by unauthorized devices. However, the flag itself would not require that all machines recognize it, and would act only as a means to halt unauthorized use in machines capable of detecting it.

In November 2003, the FCC published a Report and Order that required all digital devices capable of receiving digital broadcast over the air, and sold after July 1, 2005, to incorporate a standard content-protection technology that would recognize the broadcast video flag and limit redistribution when the flag is recognized.[8] The FCC's regulations apply the flag mark to all devices and receivers capable of receiving digital content. Such devices include, but are not limited to, televisions, computers, digital video-recorders (e.g., TiVo), and DVD players. The broadcast flag itself is optional for broadcasters, allowing them to determine how much copy-protection they wish to impose on their digital broadcast content.[9]

Because the flag does not prevent the distribution of content to non-compliant devices, a consumer who continues to use an older television set (or theoretically, a non-compliant demodulator) will still be able to receive and copy television programs in non-digital form. In addition, digital television sets made prior to July 1, 2005, will still enjoy digital content with no obstruction. In citing its support for a flag-based approach over encryption or other means, the FCC noted concerns over "the implementation costs and delays" associated with other solutions.[10]

In addition to the "compliance" requirements imposed on receiving devices, the FCC also imposed a "robustness" requirement that forces makers of consumer devices to ensure that circumvention is difficult. The standard of care adopted by the FCC was that of "an ordinary user using generally available tools or equipment."[11]

FCC Authority

The FCC derives its regulatory authority over digital television from both direct and ancillary statutory authority.

Digital Television Implementation under the Telecommunications Act of 1996

The Telecommunications Act of 1996 directed the FCC to promulgate regulations regarding the licensing of advanced television services. The act defined "advanced television services" as "television services provided using digital or other advanced technology."[12] In prescribing such regulations, the Commission was authorized to adopt such "technical and other requirements as may be necessary or appropriate to assure the quality of the signal used to provide advanced television services ... and prescribe such other regulations as may be necessary for the protection of the public interest, convenience, and necessity."[13]

Pursuant to the Telecommunications Act of 1996, the FCC has issued regulations regarding spectrum allocation for digital television stations and has established a time line for the implementation of digital broadcasting by licensees.[14] At least one court has agreed that in regard to television digital tuners, the FCC possessed reasonable authority to act, based on an "unambiguous command of an act of Congress."[15]

Copyright Protection

While copyright protection generally lies outside the scope of the FCC, the Commission may exercise jurisdiction over matters not explicitly provided for by statute if the exercise is "reasonably ancillary to the effective performance of the Commission's various responsibilities for the regulation of television broadcasting."[16] The FCC has asserted that television receivers generally, and digital television receivers specifically, fall within the scope of that authority.[17] Under the FCC Report and Order, "pursuant to the doctrine of ancillary jurisdiction, we adopt use of the ... flag as currently defined for redistribution control purposes and establish compliance and robustness rules for devices with demodulators to ensure that they respond and give effect to the ... flag." However, the FCC initially put off deciding on permanent mechanisms for approving "content protection and recording technologies to be used in conjunction with device outputs." Instead, in its Report and Order, the FCC proposed examination of such issues at a later time and established an interim certification process for currently proposed devices.[18] In addition to the need to regulate television broadcasting, the FCC's action arguably protects broadcasters from any unreasonable loss in advertising revenue that may result from unauthorized sharing of copyrighted digital television broadcasts.

However, the FCC was careful to note that the "scope of our decision does not reach existing copyright law," and that its rulemaking established a "technical protection measure" that did not change the underlying "rights and remedies available to copyright holders." In addition, "this decision is not intended to alter the defenses and penalties applicable in cases of copyright infringement, circumvention, or other applicable laws."[19]

POSSIBLE IMPLICATIONS OF THE BROADCAST VIDEO FLAG

While the broadcast flag is intended to "prevent the indiscriminate redistribution of [digital broadcast] content over the Internet or through similar means," the goal of the flag was not to impede a consumer's ability to copy or use content lawfully in the home, nor was the policy intended to "foreclose use of the Internet to send digital broadcast content where it can be adequately protected from indiscriminate redistribution."[20] However, current technological limitations have the potential to hinder some activities that might normally be considered "fair use" under existing copyright law.[21] For example, a consumer who wishes to record a program to watch at a later time, or at a different location (time-shifting and space-shifting, respectively), might be prevented when otherwise approved technologies do not allow for such activities or do not integrate well with one another, or with older, "legacy" devices. In addition, future fair or reasonable uses may be precluded by these limitations. For example, a student would be unable to e-mail herself a copy of a project with digital video content because no current secure system exists for e-mail transmission.

In addition, some consumer electronics and information technology groups contend that the licensing terms for approving new compliant devices are limiting and may potentially stifle innovation, especially with regard to computer hardware.[22]

While the FCC in its Report and Order declined to establish formal guidelines for which "objective criteria should be used to evaluate new content protection and recording technology," it has stated an intention to take up these issues in the future.[23]

Finally, consumer rights and civil liberties groups worry about the possibility that such content protections will limit the free flow of information and hamper the First Amendment. This concern is expressed most prominently regarding news or public interest-based content, or works that have already entered the public domain. Despite suggestions raised by consumer rights groups, the FCC has so far declined to adopt language to prevent content providers from using the broadcast flag on such programs, largely because of the "practical and legal difficulties of determining which types of broadcast content merit protection from indiscriminate redistribution and which do not."[24]

LEGAL CHALLENGES TO THE BROADCAST VIDEO FLAG

In October of 2004, the American Library Association (ALA), Association of Research Libraries, American Association of Law Libraries (AALL), Medical Libraries Association, and others petitioned the U.S. Court of Appeals for the District of Columbia Circuit to review the FCC's Report and Order. Bringing a challenge on behalf of "libraries, librarians and educators ... and ... television viewers and computer users," the petitioners, as parties to the agency proceedings, questioned the FCC's statutory authority to establish the broadcast flag system under the Communications Act of 1934. On May 6, 2005, the United States Court of Appeals for the District of Columbia Circuit granted the ALA's petition for review and reversed and vacated the Commission's order requiring DTV reception equipment to be manufactured with the capability to prevent unauthorized redistributions of digital content.[25]

In *American Library Association v. Federal Communications Commission*, the court of appeals determined that the FCC lacked the authority "to regulate apparatus that can receive television broadcasts when those apparatus are not engaged in the process of receiving a broadcast transmission."[26] The court noted that in adopting the broadcast flag rules, the Commission "cited no specific statutory provision giving [it] authority to regulate consumers' use of television receiver apparatus after the completion of the broadcast transmission."[27] The Commission's reliance on its ancillary jurisdiction under Title I of the Communications Act of 1934 was rejected by the court. The court found that although the jurisdictional grant under Title I plainly encompasses the regulation of apparatus that can receive television broadcast content, the Commission's regulatory authority does not extend beyond the actual receipt of such content by the apparatus in question. The court's decision was limited to resolving whether the Commission had the authority to impose the broadcast flag requirements; it did not address the imposition of the broadcast flag requirements in terms of copyright law.

BROADCAST VIDEO FLAG LEGISLATION INTRODUCED IN THE 109TH CONGRESS

In response to the *American Library Association* decision, two bills were introduced in the 109th Congress that would have expressly granted statutory authority to the FCC under the Communications Act of 1934 to implement the FCC's R*eport and Order In the Matter of Digital Broadcast Content Protection*. These legislative proposals represent approaches that may be taken in the 110th Congress for authorizing a broadcast video flag system. The Digital Content Protection Act of 2006 was introduced by Senator Ted Stevens in the 109th Congress as part of two bills, S. 2686 and H.R. 5252 (as reported in the Senate).[28]

Section 452 of S. 2686, as introduced, would have required the FCC to modify its Report and Order to permit the following transmissions:

- short excerpts of broadcast digital television content over the Internet,
- broadcast digital television content over a home network or other localized network accessible to a limited number of devices connected to such network, and
- redistribution of news and public affairs programming (not including sports) in which the primary commercial value depends on timeliness, as determined by the broadcaster or broadcasting network.

The Senate version of H.R. 5252 would have prohibited television broadcast stations from using the broadcast video flag "to limit the redistribution of news and public affairs programming the primary commercial value of which depends on timeliness." However, the bill expressly allowed each broadcaster or broadcasting network to make the determination as to whether the primary commercial value of a particular news program depends on timeliness. The bill also authorized the FCC to "review any such determination by a broadcaster or broadcasting network if it receives bona fide complaints alleging, or otherwise has reason to believe, that particular broadcast digital television content has violated" this limitation concerning timeliness and commercial value.

Hearings on the broadcast flag held by the 109th Congress revealed that educators and librarians who use digital materials in education are concerned that a broadcast video flag regime could frustrate the utilization of digital television in distance education.[29] Both bills contained provisions that sought to preserve this statutory right under the Technology, Education, and Copyright Harmonization (TEACH) Act of 2002:[30]

- S. 2686 required the FCC's video flag regulation to "permit government bodies or accredited nonprofit educational institutions to use copyrighted work in distance education courses pursuant to" the TEACH Act and the amendments made by that Act.[31]
- H.R. 5252, as reported in the Senate, contained a provision that would have directed the FCC to, within 30 days of enactment of the Act, initiate proceedings "for the approval of digital output protection technologies and recording methods for use in the course of distance learning activities."[32]

In addition, the Senate version of H.R. 5252 clarified that nothing in the bill shall "be construed to affect rights, remedies, limitations, or defenses to copyright infringement, including fair use," under the Copyright Act. S. 2686 did not contain a similar provision with regard to the broadcast video flag.[33]

REFERENCES

[1]   For information about a proposed flag for digital *audio* broadcasts, *see* CRS Report RS22489, *Copyright Protection of Digital Audio Radio Broadcasts: The "Audio Flag,"* by Jared Huber and Brian T. Yeh.

[2]   Federal Communications Commission, In the Matter of Advanced Television Systems and Their Impact Upon the Existing Television Broadcast Service: Fifth Report and Order, 12 F.C.C. Rec. 12809, 12811-12812 (1997) (hereinafter FIFTH REPORT).

[3]   This date is codified at 47 U.S.C. § 309(j)(14)(A).

[4]   For more information on DTV, see CRS Report RL31260, *Digital Television: An Overview*, by Lennard Kruger.

[5]   However, it should be noted that while duplication is fairly simple, distribution, especially of high quality digital content, can be quite time-consuming. For example, over broadband connections, while a music file may take a matter of minutes, television shows in standard analog format take a number of hours. Digital programs (such as an

hour of high-definition television programming) in turn, may take upwards of 10-15 hours to successfully download.

[6]   FIFTH REPORT at 12811 ¶ 5.

[7]   This collaboration was open to any group or individual wishing to participate, with the exception of the press. BPDG, *Final Report of the Co-Chairs of the Broadcast Protection Discussion Subgroup to the Copy Protection Technical Working Group*, FN 4, (June 3, 2002), *available on Janaury 10, 2007 at* [http://www.cptwg.org/ Assets/TEXT%20FILES/ BPDG/BPDG%20Report.DOC]).

[8]   FCC, In the Matter of Digital Broadcast Content Protection: Report and Order and Further Notice of Proposed Rulemaking, MB Docket No. 02-230, 18 F.C.C.R. 23550, 23589 (November 4, 2003) (hereinafter REPORT AND ORDER).

[9]   The amount of copy protection has a potential for variability. For instance, a content provider such as C-SPAN could decide that no copy protection is necessary and would set the flag to an off-position. Digital content would therefore be available without any restrictions to the user. However, a broadcaster who sought to show pay-per-view digital content might choose to set the flag to an on-position, which would disallow any form of copying, and would potentially add a setting to restrict the amount of time a user has to watch the program after purchase. Alternatively, a content provider may decide that individual copying is permitted, provided a user views that copy on a secure, compliant device.

[10]   REPORT AND ORDER at 23561.

[11]   REPORT AND ORDER, Appendix B, at 23592.

[12]   47 U.S.C. § 336(i)(1).

[13]   47 U.S.C. §§ 336(b)(4) and (5).

[14]   47 C.F.R. § 73.624 (2004). *See also* [http://www.fcc.gov/mb/policy/dtv/] for an overview of the FCC's activities with regard to the implementation of DTV.

[15]   Consumer Electronics Association v. FCC, 347 F.3d 291, 301 (D.C. Cir. Oct. 28, 2003).

[16]   United States v. Southwestern Cable Co., 392 U.S. 157, 178 (1968).

[17]   As to ancillary jurisdiction, *see id.* at 178. Concerning the FCC's ancillary authority, Congress has given the Commission "a comprehensive mandate," with "expansive powers," which has led the courts to conclude that the Communications Act of 1934 provides the Commission with ancillary jurisdiction over matters that are related to the provision of radio or television service, though not specifically enumerated in the act. Historically the FCC has exercised its ancillary jurisdiction to promulgate regulations in a number of areas. *See id.* at 173, 177; *U.S. v. Midwest Video Corp*, 406 U.S. 649 (1972). In addition to this historic authority, the FCC relies on the definition of "wire/radio communications," which includes "all incidental 'instrumentalities, facilities, apparatus and services' that are used for the 'receipt, forwarding, and delivery' of such transmissions" as a basis for its authority over television receivers. REPORT AND ORDER at 23563.

[18]   FCC, REPORT AND ORDER, at 23575.

[19]   *Id.* at 23555.

[20]   *Id.*

[21]   An owner of a copyright has a number of exclusive rights under the Copyright Act (17 U.S.C. § 101 *et seq.*), including the exclusive right to reproduce and distribute copies. However, this right is subject to certain statutory exceptions, including the fair use

exception (17 U.S.C. § 107). This exception "permits courts to avoid rigid application of the copyright statute when, on occasion, it would stifle the very creativity which that statute is designed to foster." Dr. Seuss Enters., L.P. v. Penguin Books USA, 109 F.3d 1394, 1399 (9th Cir. 1997).

[22] Center for Democracy and Technology, *Implications of the Broadcast Flag: A Public Interest Primer*, (December 2003), *available on January 10, 2007 at* [http://www.cdt.org/copyright/20031216broadcastflag.pdf]).

[23] REPORT AND ORDER at 23578.

[24] *Id.* at 23568-23569 (internal citation omitted).

[25] 406 F.3d 689 (D.C. Cir. 2005).

[26] *Id.* at 691.

[27] Id.

[28] H.R. 5252, the Communications Opportunity, Promotion, and Enhancement (COPE) Act of 2006, was passed by the House and was then amended in the nature of a substitute by the Senate Commerce Committee, which struck everything after the enacting clause and inserted the language of S. 2686. The House-passed version of H.R. 5252 did not contain a video flag provision.

[29] The Broadcast and Audio Flag: Hearing Before the Sen. Comm. on Commerce, Science, and Transportation, 109th Cong., 2nd Sess. (2006) (statement of Jonathan Band, counsel of the American Library Association), at 1-5, available on January 10, 2007 at [http://commerce.senate.gov/pdf/band012406.pdf].

[30] For more information on the TEACH Act, see CRS Report RL33516, Copyright Exemptions for Distance Education: 17 U.S.C. 110(2), the Technology, Education, and Copyright Harmonization Act of 2002, by Jared A. Huber, Brian T. Yeh, and Robin Jeweler.

[31] Section 452 of S. 2686 (as introduced), 109th Cong., 2d. Sess. (2006).

[32] Section 452 of H.R. 5252 (reported in the Senate), 109th Cong., 2d Sess. (2006).

[33] However, S. 2686 would have established a Digital Audio Review Board to draft a proposed regulation governing the use of an "audio" flag for digital audio broadcasts; such a regulation was to be "consistent with fair use principles."

In: Telecommunications and Media Issues
Editors: A. N. Moller, C. E. Pletson, pp. 97-102

ISBN: 978-1-60456-294-1
© 2008 Nova Science Publishers, Inc.

.

Chapter 7

WIPO Treaty on the Protection of Broadcasting Organizations*

Margaret Mikyung Lee

Abstract

Because existing international agreements relevant to broadcasting protections do not cover advancements in broadcasting technology that were not envisioned when they were concluded, in 1998 the Standing Committee on Copyright and Related Rights (SCCR) of the World Intellectual Property Organization (WIPO) decided to proceed with efforts to negotiate and draft a new treaty that would extend protection to new methods of broadcasting, but has yet to achieve consensus on a text. In recent years, a growing signal piracy problem has increased the urgency of concluding a new treaty, resulting in a decision to restrict the focus to signal-based protections for traditional broadcasting organizations and cablecasting. Consideration of controversial issues of protections for webcasting (advocated by the United States) and simulcasting will be postponed. However, considerable work remains to achieve a final proposed text as the basis for formal negotiations to conclude a treaty by the end of 2007, as projected. A concluded treaty would not take effect for the United States unless Congress enacts implementing legislation and the United States ratifies the treaty with the advice and consent of the Senate. Noting that the United States is not a party to the 1961 Rome Convention, various stakeholders have argued that a new broadcasting treaty is not needed, that any new treaty should not inhibit technological innovation or consumer use, and that Congress should exercise greater oversight over U.S. participation in the negotiations.

As part of WIPO's Digital Agenda, a WIPO Treaty on the Protection of Broadcasting Organizations is envisioned to adapt broadcasters' rights to the digital era. Broadcasting industry advocates of the need for this treaty observe that existing relevant international agreements[1] do not offer sufficient protection because advances in broadcasting technology and the parallel evolution of the industry are not covered by the terms of existing agreements.

---

* Excerpted from CRS Report RS22585, dated January 26, 2007.

These proponents note that the primary agreement covering broadcasting and cablecasting rights, the Rome Convention, was concluded in 1961 and predates home audio and video recording, telecommunications satellite systems and consumer satellite dishes, digital technology, wireless networks, and the ability of consumers to receive broadcasts via computer or mobile telephone. Accordingly, proponents assert the Convention does not adequately protect these new modes of broadcasting.

The proposed new broadcasting treaty would grant broadcasting and cablecasting organizations protection of their program transmissions for a fixed term of years, enabling them to prohibit copying and redistribution of transmissions without authorization, which could be enforced through technological means of preventing circumvention of encrypted transmissions and the like. Such protections would be distinct from the copyright of the creators of the content for program transmissions. However, opponents of the treaty respond that it is not necessary, noting that the development of the broadcasting industry in the United States has not been hurt by the fact that it is not even a party to the Rome Convention.

From its first session in November 1998, the SCCR decided to pursue in earnest discussions and submissions concerning the text of a new broadcasting treaty. Since 2004, the SCCR has been pushing for a diplomatic conference for final negotiations and adoption of a treaty; however, after eight years and fifteen sessions of preliminary negotiations, no consensus has been reached on a text adequate for a diplomatic conference. At its May 2006 meeting, the SCCR decided to drop webcasting (transmitting over the Internet) and simulcasting (transmitting simultaneously via traditional broadcasting over the air and on the Internet) from the scope of the treaty, placing them into a separate, parallel negotiating track. The United States was almost the sole proponent of including webcasting in the treaty and had tried to bolster support for it by linking it to simulcasting, which the European Union advocated. The SCCR hoped to increase the likelihood of successfully concluding the treaty by dropping these highly controversial issues.

At its fall 2006 meeting, the WIPO General Assembly tentatively agreed to convene a diplomatic conference in November/December 2007 to conclude a treaty for the protection of only traditional broadcasting organizations and cablecasting organizations, contingent on the SCCR's successfully tabling a consensus proposed text. To that end, the SCCR would hold two special sessions, one in January 2007, which has just concluded, and another in June 2007, to "aim to agree and finalize, on a signal-based approach, the objectives, specific scope and object of protection."[2] The emphasis on a signal-based approach was an attempt to narrow the focus of the treaty to signal theft and piracy in order to allay concerns that a new layer of intellectual property rights in the content of broadcasts would, in effect, extend protection beyond the expiration of copyrights for each broadcast transmission and keep or remove content from the public domain. Since reportedly significant differences yet remain among the positions of various parties, the conclusion of a treaty by the end of this year is uncertain.

The Revised Draft Basic Proposal, WIPO Doc. SCCR/15/2 (July 31, 2006)[3] remains the basis for negotiations and will be the default text if the SCCR fails to achieve a consensus text that would enable a diplomatic conference to proceed. The Revised Draft Basic Proposal was considered inadequate to support a successful diplomatic conference because it essentially incorporates every major alternative text for the articles about which there remain major differences among the WIPO parties. For example, there are two alternatives for Article 18, one providing that the term of protection shall be 50 years, the other, that the term shall be 20 years. The protections available under the Rome Convention have a term of 20 years and the longer 50-year term proposed for the new treaty has been controversial. Furthermore, this text does not define a "signal," although the Chairman of the SCCR floated a proposed definition of "signal" in an informal "non-paper" at the first special session in January 2007.[4] There appears to be uncertainty and disagreement among the negotiating parties as to precisely what a "signal-based" approach means for the narrowed focus of a new treaty. Consequently, some parties suggest that a "signal-based" approach, mandated by the WIPO General Assembly, may still encompass certain elements of exclusive rights including the right to prohibit certain uses of a broadcast, which remains a major point of contention. These two examples are indicative of the lack of consensus affecting most of the provisions of the Revised Draft Basic Proposal. Therefore, it may be useful to consider some of the major points of contention for the treaty.

The principles expressed in various stakeholder statements are fairly representative of common objections raised by treaty opponents and also of some of the concerns or positions expressed by various WIPO country-parties during negotiations. A joint statement distributed by 41 corporations, industry associations, and non-governmental organizations at the first special session of the SCCR advocated several guidelines for a treaty text, while not conceding their position that a treaty is not necessary at all. This statement is similar to earlier statements issued by many of the same stakeholders at the September 2006 meeting of the WIPO General Assembly and to positions expressed at stakeholder roundtables held by the U.S. Patent and Trademark Office (USPTO) in September 2006 and January 2007.[5] The stakeholders issuing the statements comprise a range of organizations representing Internet service providers, computer technology companies, libraries and information professionals, content creators/owners, and consumer groups.

First, the stakeholders assert there is no need for a treaty: "The United States has a flourishing and well-capitalized broadcasting and cablecasting sector, notwithstanding its decision not to accede to the [Rome Convention]. We see no necessity for the creation of new rights to stimulate economic activity in this area. [Longstanding negotiations do not] justify the creation of rights that would be exceedingly novel in U.S. law and that are likely to harm consumers' existing rights, and stifle technology innovation."[6] Before the creation of such rights, the stakeholders maintain that "there should be a demonstrated need for such rights, and a clear understanding of how they will impact the public, educators, existing copyright holders, online communications, and new Internet technologies."[7]

Second, according to the stakeholders, the treaty should not be "rights-based," that is, grant exclusive rights in broadcasts similar to copyright. Rather, it should be, in their view, "signal-based," meaning that the prevention of theft or piracy of pre-broadcast signals should be the focus of the treaty. Third, stakeholders assert that the treaty should not be negotiated with reference to whether it detracts or departs from the Rome Convention, although the signers of the statement believe that strong signal protections are consistent with the Rome Convention. The European Union in particular has advocated that a new treaty should comply with the Rome Convention. However, some stakeholders observed[8] that the narrowed treaty focus on a signal-based approach is more akin to the Brussels Convention.[9] Fourth, to the extent the treaty permits rights beyond protection against signal theft/piracy, the stakeholders claim that mandatory limitations and exceptions similar to those under copyright laws should be included in the treaty to ensure that the treaty does not prohibit uses of broadcast content that are lawful under copyright law. The treaty should also, in their view, permit additional limitations and exceptions appropriate in a digital network environment. Fifth, the stakeholders contend that the treaty should exclude coverage of fixations, transmissions or retransmissions over a home network or personal network.

Concerns have been raised that because the Revised Draft Basic Proposal envisions protections for technological protections measures (TPM) and digital rights management schemes (DRM), the beneficiary broadcasting organizations would have the ability to control signals in a home or personal network environment. Stakeholders allege that this would inhibit such networking services and related technology innovations. Sixth, despite the removal of webcasting and simulcasting from the scope of the treaty, the phrase "by any means" in various articles of the Revised Draft Basic Proposal would, in the stakeholders' view, include control over Internet retransmissions of broadcasts and cablecasts. Finally, to the extent that Internet transmissions may be included in the scope of the treaty, stakeholders advocate that it should ensure that intermediate network service providers are not subject to liability for alleged infringement of rights or violations of prohibitions due to actions in the normal course of business or actions of customers.

The South Centre, an intergovernmental organization of developing countries, issued a research paper on the broadcast treaty in September, 2006, which expressed some of the same concerns with regard to the benefits that the treaty would have for developing countries, as well as additional concerns.[10] Recommendations similar to those discussed above include that the negotiators: (1) consider maintaining that the rationale and scope of application of the new instrument be limited to signal protection; (2) do not accept the inclusion of any exclusive rights, or at the least, that such rights do not extend beyond those incorporated in the Rome Convention, unless clear evidence is found for the need to grant such rights and mechanisms to address the potential harms they may cause are developed; and (3) ensure that appropriate safeguards to pursue public policy objectives and limitations and exceptions are included in the text. Additionally, the South Centre recommends that the negotiators: (1) refrain from expanding protection to include delivery via computer networks as well as any reference to webcasting (which is at odds with the position of the United States and webcasting advocates); (2) provide for special treatment to public service broadcasting and/or discrimination between commercial and non-commercial broadcasting; (3) limit the maximum term of protection to 20 years, if exclusive rights are required for signal protection, rather than the 50 years in the Revised Draft Basic Proposal; and (4) do not include obligations concerning the protection of TPMs and DRM schemes, or at least consider including limitations and exceptions as minimum standards to these obligations to ensure they do not impede access to content.

As noted above, the United States has been the primary advocate for extending protections to webcasting, whether in a new broadcasting treaty or in a separate agreement or protocol. In a statement submitted to the SCCR, the United States clarified that it "never intended that protection be afforded to the ordinary use of the Internet or World Wide Web, such as through e-mail, blogs, websites and the like. We intended only to cover programming and signals which are like traditional broadcasting and cablecasting, i.e. simultaneous transmission of scheduled programming for reception by the public."[11] In the statement, the United States sought to replace the term "webcasting" with "netcasting" and clarified that "netcasting" was limited to transmissions over computer networks carrying programs consisting of audio, visual or audio-visual content or representations thereof which are of the type that can be, but are not necessarily, carried by the program carrying signal of a broadcast or cablecast, and which are delivered to the public in a format similar to broadcasting or cablecasting. It decided that "webcasting" "unnecessarily implied that ordinary activity on the World Wide Web would be covered by the definition." The United States affirmed its advocacy of extending the same protections to "netcasting" as were and would be extended to traditional broadcasting and cablecasting, but asserted that such protections would only be whatever was necessary to prevent signal theft/piracy.

Assuming that the treaty is eventually successfully concluded and that the United States is a signatory, any such treaty would not take effect for the United States unless and until the treaty was ratified by the United States with the advice and consent of the Senate, and Congress enacted implementing legislation. Furthermore, if the final text of the treaty adopted by WIPO includes Alternative AAA to Article 27 of the Revised Draft Basic Proposal, a party to the new broadcast treaty would be required to become a party to the Rome Convention first, which would mean that the United States would also have to consider ratification of that Convention, to which it is not currently a party.[12] Implementing legislation would likely be necessary to establish new protections or amend existing ones in broadcasting laws and perhaps copyright laws. Currently, 47 USC §§ 325 and 605 and 18 USC §§ 2510-2512 provide for broadcasting protections and title 17 of the U.S. Code contains the copyright laws. Additionally, webcasting/netcasting and simulcasting may be included in a separate agreement or as a protocol to a new broadcasting treaty, unless they are reconsidered for inclusion in the new broadcast treaty itself.

Certain stakeholders that are either opposed to the treaty or concerned about the inclusion of certain protections have called on Congress to hold hearings on the treaty to determine whether a new treaty is necessary or at least to exercise greater oversight over the U.S. delegation's positions on the treaty.[13] They had also urged that the U.S. Copyright Office and the USPTO solicit public commentary, which those agencies did through the aforementioned roundtables. These stakeholders are concerned that without public input, major changes in U.S. telecommunications and copyright laws will be effected via implementation of a new broadcast treaty without a full opportunity for domestic debate.[14] Partly in response to the objections raised by stakeholders in the information and communications technology industries, the United States reportedly sought to ensure that a diplomatic conference would not proceed if special sessions failed to resolve the major disagreements.[15]

REFERENCES

[1]   Rome Convention for the Protection of Performers, Producers of Phonograms and Broadcasting Organizations (Rome Convention), the Trade-Related Intellectual Property Rights Agreement of the World Trade Organization (WTO TRIPS) and the Brussels Convention Relating to the Distribution of Program-Carrying Signals Transmitted by Satellite (Brussels Convention).

[2]   *Report*, WIPO Doc. WO/GA/33/10 (Oct. 3, 2006).

[3]   Available at [http://www.wipo.int/edocs/mdocs/sccr/en/sccr_15/sccr_15_2.pdf] (last visited Jan. 25, 2007).

[4]   Text reprinted at [http://www.ip-watch.org/weblog/index.php?p=508 and res=1024 and print=0] (last visited Jan. 25, 2007). "Signal" would be defined as "an electronically-generated carrier capable of transmitting programmes."

[5]   Links to these statements are accessible via [http://www.eff.org/IP/WIPO/ broadcasting_treaty/] (last visited Jan. 25, 2007).

[6]   Statement of Electronic Frontier Foundation to USPTO Roundtable on Proposed WIPO Broadcasting Treaty, Sept. 5, 2006, [http://www.eff.org/IP/WIPO/ broadcasting_treaty/ EFF_uspto _090506.pdf] (last visited Jan. 25, 2007).

[7]   Id.

[8]   William New, *WIPO Negotiators Try to Bear Down on Broadcasting Treaty*, Intellectual Property Watch (Jan. 18, 2007), available at [http://www.ip-watch.org/weblog/index.php?p= 509 and res=1024 and print=0] (last visited Jan. 25, 2007).

[9]   The Convention provides for the obligation of each contracting State to take adequate measures to prevent the unauthorized distribution on or from its territory of any program-carrying signal transmitted by satellite. The distribution is unauthorized if it has not been authorized by the organization — typically a broadcasting organization — that has decided what the program consists of. The obligation applies to organizations that are nationals of a Convention party. However, the Convention provisions are not applicable where the distribution of signals is made from a direct broadcasting satellite.

[10] Viviana Munoz Tellez and Andrew Chege Waitara, South Centre Research Paper 9, The Proposed WIPO Treaty on the Protection of Broadcasting Organisations: Are New Rights Warranted and Will Developing Countries Benefit? (September 2006), available at [http://www.southcentre.org/publications/researchpapers/ResearchPapers9.pdf] (last visited Jan. 25, 2007).

[11] Submission of the United States of America to the WIPO Standing Committee on Copyright and Related Rights, WIPO Doc. SCCR/15/INF/2 (Aug. 22, 2006).

[12] It was also not a signatory when the Convention was concluded, so it would appear that Congress has never previously considered the Convention.

[13] Examples of such letters are available via [http://www.eff.org/IP/ WIPO/broadcasting_treaty/] (last visited Jan. 25, 2007).

[14] IT, Consumer Groups Question Need for New Broadcasters' Treaty, 20 World Intellectual Property Report (BNA, October 2006).

[15] William New, *Agreement Reached on WIPO Development Agenda, Patents; No Broadcasting Yet*, Intellectual Property Watch (Sept. 30, 2006), available at [http://www.ip-watch.org/weblog/ index.php?p=410 and res=1024 and print=0] (last visited Jan. 25, 2007).

In: Telecommunications and Media Issues
Editors: A. N. Moller, C. E. Pletson, pp. 103-109

ISBN: 978-1-60456-294-1
© 2008 Nova Science Publishers, Inc.

Chapter 8

COPYRIGHT PROTECTION OF DIGITAL AUDIO RADIO BROADCASTS: THE "AUDIO FLAG"*

Jared Huber and Brian T. Yeh

ABSTRACT

Protecting audio content broadcasted by digital and satellite radios from unauthorized dissemination and reproduction is a priority for producers and owners of those copyrighted works. One technological measure that has been discussed is the Audio Protection Flag (APF or "audio flag"). The audio flag is a special signal that would be imbedded into digital audio radio transmissions, permitting only authorized devices to play back copyrighted audio transmissions or allowing only limited copying and retention of the content. Several bills introduced in the 109th Congress would have granted the Federal Communications Commission (FCC) authority to promulgate regulations to implement the audio flag. The parties most likely affected by any audio flag regime (including music copyright owners, digital radio broadcasters, stereo equipment manufacturers, and consumers) are divided as to the anticipated degree and scope of the impact that a government-mandated copyright protection scheme would have on the "fair use" rights of consumers to engage in private, noncommercial home recording. Critics of the audio flag proposal are concerned about its effect on technological innovation. However, proponents of the audio flag feel that such digital rights management (DRM) technology is needed to thwart piracy or infringement of intellectual property rights in music, sports commentary and coverage, and other types of copyrighted content that is transmitted to the public by emerging high-definition digital radio services (HD Radio) and satellite radio broadcasters.

This report provides a brief explanation of the audio flag and its relationship to digital audio radio broadcasts, and summarizes legislative proposals considered by the 109th Congress, including H.R. 4861 (Audio Broadcast Flag Licensing Act of 2006) and S. 2686 (Digital Content Protection Act of 2006), that would have authorized its adoption. Although not enacted, these two bills represent approaches that may be taken in the 110th Congress to authorize the use of an audio flag for protecting broadcast digital audio content.

---

* Excerpted from CRS Report RS22489, dated January 16, 2007.

INTRODUCTION

Although the advent of digital technology has brought about higher quality for audio and video content, creators of such content and policy makers are concerned that, without adequate content protection measures, unlawful digital copying and distribution of copyrighted material may endanger the viability of the motion picture, television, and music industries.[1] As a result, technological measures have been proposed that are aimed at protecting copyrighted media from, among other things, unauthorized reproduction, distribution, and performance. One of these content protection schemes is the Audio Protection Flag (APF or "audio flag"), which would protect the content of digital radio transmissions against unauthorized dissemination and reproduction.

BACKGROUND

What Is Digital Audio?

To understand digital audio, an explanation of how analog and digital technology differ is helpful. Analog technology is characterized by an output system where the signal output is always proportional to the signal input. Because the outputs are analogous, the word "analog" is used. An analog mechanism is one where data is represented by continuously variable physical quantities like sound waves or electricity. Analog audio technologies include traditional radio (AM/FM radio), audio cassettes, and vinyl record albums. These technologies may deliver imprecise signals and background noise. Thus, the duplication of analog audio often erodes in quality over time or through "serial copying" (the making of a copy from copies).

The term "digital" derives from the word "digit," as in a counting device. Digital audio technologies represent audio data in a "binary" fashion (using 1s and 0s). Rather than using a physical quantity, a digital audio signal employs an informational stream of code. Consequently, the code from a digital audio source can be played back or duplicated nearly infinitely and without any degradation of quality. Digital audio technologies include digital radio broadcasts (such as high-definition radio, or "HD Radio"), satellite radio, Internet radio, compact discs, and MP3-format music files.

Digital Content Protection

With the advent of digital technology, content providers have been interested in using content security measures to prevent unauthorized distribution and reproduction of copyrighted works. These technology-based measures are generally referred to as "digital rights management," or DRM. As the name suggests, DRM applies only to digital media (which would include analog transmissions converted into digital format). Examples of DRM include Internet video streaming protections, encrypted transmissions, and Content Scrambling Systems (CSS) on DVD media.

In 1998, Congress passed the Digital Millennium Copyright Act (DMCA). The DMCA added a new chapter 12 to the Copyright Act, 17 U.S.C. §§ 1201-1205, entitled "Copyright Protection and Management Systems." Section 1201(a)(1) prohibits any person from circumventing a technological measure that effectively controls access to a copyrighted work. This newly created right of "access" granted to copyright holders makes the act of gaining access to copyrighted material by circumventing DRM security measures, itself, a violation of the Copyright Act. Prohibited conduct includes descrambling a scrambled work, decrypting an encrypted work, and avoiding, bypassing, removing, deactivating, or impairing a technological measure without the authority of the copyright owner.[2] In addition, the DMCA prohibits the selling of products or services that circumvent access-control measures, as well as trafficking in devices that circumvent "technological measures" protecting "a right" of the copyright owner.[3]

In contrast to copyright infringement, which concerns the unauthorized or unexcused use of copyrighted material, the anti-circumvention provisions of the Copyright Act prohibit the design, manufacture, import, offer to the public, or trafficking in technology produced to circumvent copyright encryption programs, regardless of the actual existence or absence of copyright infringement.

THE "AUDIO FLAG"

One form of DRM technology that may be used to protect the content of digital audio transmissions from unauthorized distribution and reproduction is the "audio flag." The flag has two primary aspects: a physical component and rules and standards that define how devices communicate with flagged content transmitted from digital audio sources. For instance, a satellite digital audio radio stream of a particular broadcast music program could contain an audio flag (the mechanism) that prohibits any reproduction or further dissemination of the broadcast (the standard). The audio flag, according to its proponents, would operate in a similar manner as the broadcast video flag that has been proposed for digital television transmissions.[4] Functionally, the audio flag system would work by embedding a special signal within transmitted digital audio data, informing the receiving device of certain copyright restrictions on the use of the content by the listener — for example, limiting the number of copies of a recording that the user may make.

Those advocating the use of an audio flag for digital radio programming include musicians, songwriters, record labels, and other providers of audio content that could be broadcast to the public through digital transmissions. The Copyright Act bestows several exclusive rights upon the creator of a work (or the individual having a legal interest in the work) that permit the copyright holder to control the use of the protected material. These statutory rights allow a copyright holder to do or to authorize, among other things, reproducing the work, distributing copies or phonorecords of the work, and publicly performing the work.[5] Parties holding a copyright interest in content transmitted through digital radio services are interested in ensuring that such content is protected from unauthorized reproduction and distribution by the broadcast recipient; the audio flag, in their view, is an effective way to achieve this objective and enforce their rights.

Proponents of audio flag technology also suggest that it would help prevent certain digital radio services (like satellite radio) from becoming a music download service through the creation of recording and storage devices that allow for further reproduction and distribution of audio broadcasts.[6] Some copyright holders argue that these broadcasters must either pay additional royalties for the privilege of offering what appears to be a music download service, or comply with an audio flag regime that will effectively prevent broadcasters from allowing the recording in the first place.[7]

RIGHTS THAT MAY BE AFFECTED BY THE AUDIO FLAG PROPOSAL

Critics of the audio flag proposal raise concerns that such a government-mandated measure may stifle technological innovation and restrict the rights of consumers to record broadcast radio — conduct that, according to audio flag opponents, is protected by the Audio Home Recording Act of 1992, as well as "fair use" principles in copyright law.

The Audio Home Recording Act of 1992

The introduction of the Digital Audio Tape (DAT) by Sony and Philips in the mid-1980s prompted passage of the Audio Home Recording Act (AHRA) in 1992.[8] A DAT recorder can record CD-quality sound onto a specialized digital cassette tape. Through the Recording Industry Association of America (RIAA), sound recording copyright holders turned to Congress for legislation in response to this technology, fearing that a consumer's ability to make near-perfect digital copies of music would displace sales of sound recordings in the marketplace.[9]

The AHRA requires manufacturers of certain types of digital audio recording devices to incorporate into each device copyright protection technology — a form of DRM called the Serial Copying Management System (SCMS), which allows the copying of an original digital work, but prevents "serial copying" (making a copy from a copy). In exchange, the AHRA *exempts* consumers from copyright infringement liability for private, noncommercial home recordings of music for personal use. Manufacturers of audio equipment, sellers of digital recording devices, and marketers of blank recordable media are also protected from contributory infringement liability upon payment of a statutory royalty fee — royalties that are distributed to the music industry. Opponents of the audio flag contend that the AHRA created a "right" for consumers to make digital recordings, a practice that might be limited or even effectively revoked by audio flag mandates.[10]

Fair Use

The doctrine of "fair use" in copyright law recognizes the right of the public to make reasonable use of copyrighted material, in certain instances, without the copyright holder's consent. Because the language of the fair use statute is illustrative, determinations of fair use are often difficult to make in advance — it calls for a "case-by-case" analysis by the courts.[11]   However, the statute recognizes fair use "for purposes such as criticism, comment, news reporting, teaching, scholarship, or research."[12]   A determination of fair use by a court considers four factors: (1) the purpose and character of the use, including whether such use is of a commercial nature or is for nonprofit educational purposes; (2) the nature of the copyrighted work; (3) the amount and substantiality of the portion used in relation to the copyrighted work as a whole; and (4) the effect of the use upon the potential market for or value of the copyrighted work.[13]

In the context of digital music downloads and transmissions, some alleged copyright infringers have attempted to use the doctrine of fair use to avoid liability for activities such as sampling,[14] "space shifting,"[15] and peer-to-peer (P2P) file sharing.[16]   These attempts have not been very successful, as several federal appellate courts have ruled against the applicability of the fair use doctrine for these purposes.[17]   No litigation has yet settled the extent to which home recording of an audio broadcast (whether transmitted through digital or analog means) is a legitimate fair use.

Critics of the audio flag also suggest that it places technological, financial, and regulatory burdens that may stifle the innovation behind digital audio technologies. They argue that the audio flag may limit the functionality of digital audio transmissions in favor of analog transmissions — thereby negatively affecting the digital audio marketplace.[18]

AUDIO FLAG LEGISLATION INTRODUCED IN THE 109TH CONGRESS

Legislation that expressly delegates authority to the FCC to mandate audio flags for digital radio transmissions would appear to be necessary before the FCC could take such steps, in the wake of a decision by the U.S. Court of Appeals for the District of Columbia Circuit in 2005 that vacated an FCC order requiring digital *televisions* to be manufactured with the capability to prevent unauthorized redistributions of digital *video* content. The court ruled that the FCC lacked the statutory authority to establish such a broadcast video flag system for digital television under the Communications Act of 1934.[19]   Two bills were introduced in the 109th Congress that would have delegated such authority; these may represent legislative approaches that could be taken in the 110th Congress.

H.R. 4861, the Audio Broadcast Flag Licensing Act of 2006

This bill would have empowered the FCC to promulgate regulations governing the licensing of "all technologies necessary to make transmission and reception devices" for digital broadcast and satellite radio.[20]   The bill directed that any such digital audio regulation shall prohibit unauthorized copying and redistribution of transmitted content through the use of a broadcast flag or similar technology, "in a manner generally consistent with the purposes of other applicable law."

S. 2686, the Digital Content Protection Act of 2006

Title IV, Subtitle C of S. 2686 would have granted the FCC the authority to issue "regulations governing the indiscriminate redistribution of audio content with respect to — digital radio broadcasts, satellite digital radio transmissions, and digital radios."[21] It also directed the FCC to establish an advisory committee known as the "Digital Audio Review Board," composed of representatives from several industries, including information technology, software, consumer electronics, radio and satellite broadcasting, audio recording, music publishing, performing rights societies, and public interest groups. The Board would have been responsible for drafting a proposed regulation that reflects a consensus of the members of the Board and that is "consistent with fair use principles," although the bill did not define whether such "fair use" has the same connotation as that used in the copyright law.

REFERENCES

[1]   See Internet Streaming of Radio Broadcasts: Balancing the Interests of Sound Recording Copyright Owners with Those of Broadcasters: Hearing Before the House Subcomm. on Courts, the Internet, and Intellectual Property, 108th Cong., 2d Sess. (2004)(statement of David O. Carson, General Counsel for the U.S. Copyright Office), at 34, available on January 10, 2007 at [http://www.copyright.gov/docs/ carson 071504.pdf] ("In the absence of corrective action, the rollout of digital radio and the technological devices that promise to enable consumers to gain free access at will to any and all music they want will pose an unacceptable risk to the survival of what has been a thriving music industry. . . "); The Audio and Video Flags: Can Content Protection and Technological Innovation Coexist?: Hearing Before the House Subcommittee on Telecommunications and the Internet, 109th Cong., 2nd Sess. (2006)(statement of Mitch Bainwol, Chairman and CEO of the Recording Industry Association of America) ("[T]he music industry has faced an immense challenge in online piracy over the past several years. In addition to sharply declining sales figures, composers, artists, musicians, technicians, and a multitude of others engaged in the music industry have seen their jobs disappear.").

[2]   17 U.S.C. § 1201 (a)(3).

[3]   Id. §§ 1201(a)(2), (b).

[4]   See CRS Report RL33797, Copyright Protection of Digital Television: The Broadcast Video Flag, by Brian T. Yeh. The broadcast video flag is an embedded signal in digital television broadcasts that prohibits unauthorized redistribution of broadcast programs. See also, CRS Report RL31260, Digital Television: An Overview, by Lennard G. Kruger.

[5]   17 U.S.C. § 106.

[6]   For detailed information about satellite radio and music licensing matters, see CRS Report RL33538, Satellite Digital Audio Radio Services and Copyright Law Issues, by Brian T. Yeh.

[7]   See RIAA's Executive Comments to the FCC on HD Radio, available on January 10, 2007, at [http://www.riaa.com/news/newsletter/061604_2.asp].

[8]   P.L. 102-563 (1992), codified at 17 U.S.C. §§ 1001 et seq.

[9]   H.REPT. 102-873, at 18-19 (1992).

[10]   See, e.g., The Audio and Video Flags: Can Content Protection and Technological Innovation Coexist?: Hearing Before the House Subcommittee on Telecommunications and the Internet, 109th Cong., 2nd Sess. (2006)(statement of Gary J. Shapiro, for the Consumer Electronics Association and the Home Recording Rights Coalition), at 8-9.

[11]   Campbell v. Acuff-Rose Music, Inc., 510 U.S. 569, 577 (1994).

[12]   17 U.S.C. § 107.

[13]   17 U.S.C. §§ 107(1)-(4).

[14]   In the digital music context, sampling is a term that refers to the supposed ability of a user to make copies of copyrighted materials prior to purchase. See A&M Records, Inc. v. Napster, Inc., 114 F.Supp. 896 (N.D. Cal. 2000), aff'd in relevant part, 239 F.3d at 1018 (9th Cir. 2001).

[15]   Id. Space shifting occurs when users access CD sound recordings on their computers and portable audio devices.

[16]   P2P file sharing is facilitated by software that establishes network connections between computers to enable the exchange of data over the Internet. For more information on this topic, see CRS Report RL31998, File Sharing Software and Copyright Infringement: Metro-Goldwyn-Mayer Studios, Inc. v. Grokster, Ltd., by Brian T. Yeh and Robin Jeweler.

[17]   See, e.g., A&M Records, Inc. v. Napster, Inc., 114 F.Supp.2d 896 (N.D. Cal. 2000), aff'd, 239 F.3d 1004 (9th Cir. 2001); In re: Aimster Copyright Litigation, 334 F.3d 643 (7th Cir. 2003), cert. denied, 540 U.S. 1107 (2004).

[18]   See, e.g., The Audio and Video Flags: Can Content Protection and Technological Innovation Coexist?: Hearing Before the House Subcommittee on Telecommunications and the Internet, 109th Cong., 2nd Sess. (2006)(statement of Gigi Sohn, President of Public Knowledge) at 5-6.

[19]   Am. Library Ass'n v. FCC, 406 F.3d 689 (D.C. Cir. 2005).

[20]   H.R. 4861, 109th Cong., 2d Sess. (2006)(bill as introduced).

[21]   S. 2686, 109th Cong., 2d Sess. (2006)(bill as introduced). H.R. 5252, the Communications Opportunity, Promotion, and Enhancement (COPE) Act of 2006, was passed by the House, and then was amended in the nature of a substitute by the Senate Commerce Committee, which struck everything after the enacting clause and inserted the language of S. 2686. The House-passed version of H.R. 5252 did not contain an audio flag provision.

INDEX

D

E

F

G

T